Tapping the Markets

DIRECTIONS IN DEVELOPMENT
Private Sector Development

Tapping the Markets

Opportunities for Domestic Investments in Water and Sanitation for the Poor

Jemima Sy and Robert Warner, with Jane Jamieson

THE WORLD BANK
Washington, D.C.

Contents

Box

Figures

Tables

Foreword

Across the developing world, millions of people rely on the private sector for their daily water and sanitation needs. In the majority of cases, the providers of these essential services are not the large multinational corporations often associated with private participation in the water sector. They are local entrepreneurs operating on a small scale, who see selling water and sanitation services to the poor as market opportunities like any other.

These "base-of-the-pyramid" markets grow out of exclusion. They are the markets of the unserved—people that public services have failed to provide for and for whom internationally recognized notions of improved services are out of reach—leading them to look to self-supply. Such markets include the 1 billion people who still practice open defecation, the 2.5 billion people who use shared or unimproved sanitation facilities, and the 768 million people who use an unimproved source for drinking water. These markets will suffer the worst adverse impacts on water availability arising from the growing demand for water and climate shifts. They are at the core of the world's development challenge and the World Bank Group's core mission.

The paradox is that this large market is dominated by small, local enterprises. Once viewed as opportunists profiting from people's most basic needs, these private sector enterprises are now recognized as offering valuable services to the poor—services that would otherwise have been out of reach. Domestic private entrepreneurs are increasingly being seen as part of the solution to increasing access to water and sanitation for the poor and to enabling the poor to cope with a changing reality.

The World Bank Group commissioned this study to understand these businesses, their customers, and the factors that influence their business decisions. This understanding is critical if the Bank is to support the scale-up of safe, affordable water and sanitation services.

One of the most striking findings of this book is the sheer scale of the market potential. With respect to the countries studied here, about 20 million people are projected to get their water from rural piped water schemes managed by the private sector by 2025—that is 10 times the current number of customers, a market worth at least $90 million a year. In sanitation, the picture is the same: the current market potential for improved on-site sanitation services is estimated to be worth $2.6 billion. Behind these numbers are the hundreds of thousands of the poor and excluded who are not able to benefit from the convenience and safety of improved services.

How to service this demand and scale up access through the domestic private sector is not straightforward and is highly dependent on the country context. It will take action on all fronts, from developing toilets with more consumer appeal to simplifying licensing procedures and developing more appropriate financial products. Partnerships will be needed to support businesses and governments in piecing together the puzzle of access for the poor.

Jose Luis Irigoyen   Laurence Carter
Director   *Director*
Transport, Water, and Information and   *Advisory Services in Public-Private*
Communication Technologies   *Partnerships*
The World Bank   *International Finance Corporation*

Acknowledgments

The global study upon which this book was based was prepared by a team from the World Bank made up of staff from the Water and Sanitation Program (WSP)—a multidonor partnership administered by the World Bank—and the International Finance Corporation (IFC).

This two-part report was written by Jemima Sy (WSP) and Robert Warner (Crawford School of Public Policy, Australian National University). Jemima Sy and Jane Jamieson (IFC) served as task team leaders. Study team members included Maria Ella Lazarte and Khoi Kim Dang (WSP). The report was prepared under the direction of Jaehyang So, manager of the WSP, and Jose Luis Irigoyen, director of the World Bank's Transport, Water, and ICT (Information and Communication Technologies) Department.

Peer reviewers within the World Bank Group included William Llewelyn Davies (IFC, Sub-Saharan Africa Sustainable Business Advisory Services); Jeffrey John Delmon (World Bank, Africa Finance and Private Sector Development); Syed A. Mahmood (IFC, Investment Climate Business Regulation); Philippe Marin (World Bank, Middle East and North Africa Agriculture, Rural and Water); and Adriana de Aguinaga de Vellutini (IFC, Public-Private Partnership Advisory Services). External peer reviewers for the water study included Thierry Barbotte (Vergnet Hydro France); Brice Cabibel (Suez Environment Company); Adrien Couton (Dalberg); and Greg Koch (Global Water Stewardship, Coca-Cola Company). External peer reviewers for the sanitation study included Amy Lin (Deloitte-Monitor, Mumbai); James MacHale (American Standard); Rajiv Pradhan (International Development Enterprises, Dhaka); and Jan-Willem Rosenboom and Roshan Shrestha (Bill & Melinda Gates Foundation).

The water study was based on data and reports from three countries. Study country team members were Abdul Motaleb (WSP) and consultants from DevConsultants Limited in Bangladesh; Sylvain Adopko Migan (WSP) and consultants from Hydroconseil in Benin; and Susanna Smets, Doung Kakada, and Phyrum Kov (WSP) and consultants from GRET in Cambodia.

The sanitation study was based on data and reports from four countries. Study country team members were Rokeya Ahmed (WSP) and consultants from DevConsultants Limited in Bangladesh; Ari Kamasan (WSP) and consultants from Akademika in Indonesia; Malva Baskovich (WSP) and consultants from IMASEN in Peru; and Kaposo Boniface Mwambuli and Jason Cardosi (WSP) and consultants from PATH in Tanzania.

Editorial support was provided by Barbara Karni (copyeditor) and Kara Watkins (production manager).

About the Authors

Jemima Sy, senior water and sanitation specialist, is the global business area leader for the Water and Sanitation Program's initiative on domestic private sector participation. The Water and Sanitation Program is a multidonor partnership administered by the World Bank to support poor people in obtaining affordable, safe, and sustainable access to water and sanitation services. Jemima oversees private sector development initiatives in 16 countries, where multidisciplinary teams work on three main areas of support: public-private partnerships, market finance, and development of sanitation and water markets. A lawyer by training, Jemima has more than 10 years' experience managing water and sanitation initiatives through local private sector participation. She advises governments on regulatory reforms and privately driven performance improvement programs and is involved in developing long-term commercial financing for water and sanitation in Asia (Cambodia, Indonesia, and the Philippines) and Africa. Jemima worked with the Kenyan water regulator on the first shadow credit rating exercise of water utilities. Most recently, her work has focused on opportunities for investments in water and sanitation by small and medium-size enterprises.

Robert Warner is the director of the Pacific Research Partnerships at the Crawford School of Public Policy at the Australian National University in Canberra. Before joining the School, he was a director of the Centre for International Economics, a private consulting firm, where he undertook analyses and provided advice on economic policy and development assistance issues for clients in Australia, Africa, Asia, the Pacific, and the Caribbean. Robert spent six years at the World Bank as a country economist working on Nigeria and Ghana. Before joining the Bank, he worked at the Australian Industries Assistance Commission (now the Productivity Commission), a government agency providing advice on microeconomic policy. His main areas of interest are trade and integration, public finance, regulatory reform, and economic governance.

Jane Jamieson is a senior water industry specialist in the International Finance Corporation's Public-Private Partnerships Advisory Services department, where she undertakes public-private partnership transactions and business development—primarily in the water sector in Africa. Jane is a chartered civil engineer. Previously, in the United Kingdom, she led the Department for International Development's support to private sector participation in infrastructure and acted as infrastructure adviser to country programs in Central Asia, China, and Guyana.

Abbreviations

DPHE	Department of Public Health and Engineering
GDP	gross domestic product
GNI	gross national income
KWh	kilowatt hour
MIME	Ministry of Industry, Mining and Energy
NGO	nongovernmental organization
R&D	research and development
UNICEF	United Nations Children's Fund
WHO	World Health Organization
WSP	Water and Sanitation Program

All dollar amounts are U.S. dollars, unless otherwise indicated.

Overview

One person's challenge is another person's opportunity. Improving access to sanitation and water services is key to improving people's quality of life. Developing country governments and the international development community are looking for ways to accelerate access to improved water and sanitation services beyond the Millennium Development Goal (MDG) targets. The MDG water target will be met by 2015, and progress is being made on sanitation, but there will still be major challenges to deal with. One source of the challenge is sheer numbers: 1 billion people still practice open defecation, 2.5 billion people use shared sanitation facilities or facilities that do not meet minimum standards of hygiene, and 768 million people use an unimproved source for drinking water. Another is location: the largest gains in access to water over the last two decades have been in urban areas, where 82 percent of the population has access to water piped to their premises. Among people without access to improved water, 89 percent live in rural areas. Ninety percent of open defecation is associated with rural areas, as is 70 percent of unimproved sanitation.

Countries do not have the capacity to meet the need for improved water supplies and sanitation services from public resources alone. These challenges present an opportunity for domestic enterprises in these growing markets. Indeed, millions of poor and nonpoor households rely on the private sector to meet their needs. The range of private sector services provided goes far beyond final service delivery. In water, the private sector includes enterprises that provide water through independent systems or the resale of water. In sanitation, it includes businesses involved in the installation of latrines and toilets, the manufacture of components, the importation and sale of materials, and the provision of emptying services.

Once viewed as "opportunists" and "gap fillers," the domestic private sector is increasingly being viewed as a central part of the solution. More and more governments are interested in engaging with the private sector to increase access of the poor to services. Figuring out how best to scale up access through the domestic private sector requires an understanding of the market potential, the state of entrepreneurs' operations, and factors that shape their business environment and investment decisions.

This book examines private sector provision of piped water services and on-site sanitation services in rural areas and small towns. For rural piped water, it assesses enterprises in Bangladesh, Benin, and Cambodia; for on-site sanitation, it assesses enterprises in Bangladesh, Indonesia, Peru, and Tanzania. In all of the case study countries, the local private sector already plays a major role in providing these services. In all of them, the Water and Sanitation Program (WSP)—a multidonor partnership administered by the World Bank to support poor people in obtaining affordable, safe, and sustainable access to water and sanitation services—is actively supporting client governments in engaging the domestic private sector and is therefore well placed to offer practical follow-up of the study results.

In each country, the study examines the preferences and circumstances of poor households and the performance of enterprises that provide services directly to them. It examines commercial and investment climate factors that may affect enterprises' actual or perceived costs and risks, driving their decisions about increasing investment in their business. Specifically, the study seeks answers to the following questions:

- Is lack of interest by the domestic private sector a rational response to weak market potential, or are lack of enterprise viability and the use of inappropriate business models preventing it from taking advantage of market opportunities?
- Are investment climate factors increasing the (actual or perceived) costs and risks associated with doing business?

The report discusses opportunities and constraints for the domestic private sector and offers suggestions for addressing the constraints. Some highlights from the study are subsequently summarized.

Market Potential Is Sufficiently Large

For both piped water and on-site sanitation services, the potential market is large and expected to grow. The combined annual water sales in Bangladesh, Cambodia, and Benin through rural piped systems are expected to increase to $90 million by 2025, demonstrating the presence of demand for paid-for water. Nearly 20 million people—10 times the existing number of customers—in Bangladesh, Benin, and Cambodia could gain water access by 2025 through the private sector. In sanitation, the current market for improved on-site sanitation services is vast: untapped households represent a market of about $2.6 billion, including about $700 million in sales to poor people.

The Poor Are Willing to Pay for Value—but Not for Inferior Products and Services

Poor households are highly discriminating clients: because money is tight and incomes seasonal, they actively engage in price-value trade-offs in water and sanitation. A minimum level of water consumption from piped networks involves cash outlays that are a significant percentage of poor households' income. Most households have access to inexpensive alternative sources of water (if only for parts of the year), including wells, springs, and boreholes; therefore, they are able to dynamically manage their water demand. Poor people's purchases are limited by cost and by their assessment of the value of network water with respect to alternatives. Although poor households seem to prefer cheaper water to good-quality water, they also value the convenience of piped water. If operators can ensure good-quality service, the availability and opportunity cost of alternatives will likely shift incentives in favor of networks.

Nearly 200 million people in Bangladesh, Indonesia, Peru, and Tanzania alone have unsatisfied sanitation aspirations. Sanitation is a relatively low-priority expenditure for poor households, and cost is an important factor in their decision making. But cost is not necessarily an insurmountable barrier—as the widespread use of other consumer products, such as cellphones, by these households suggests. Poor households are willing to incur a cost to obtain attractive products and services. The problem is that the sanitation options currently on offer are not particularly appealing.

The Market Is Dominated by Small Enterprises That Are Financially Viable but Find It Hard to Scale

The base of the pyramid in the water and sanitation markets is dominated by micro- and small enterprises Most of these enterprises are able to make a profit, but they face many constraints in expanding. For water supply enterprises in Benin and Bangladesh, opportunities are limited by the public programs and policies under which these enterprises currently operate; in Cambodia, inadequate access to investment financing and the lack of security to operate as going concerns are key barriers. In sanitation, the risk of unsteady demand and the inability of small enterprises to invest in research and development and marketing limit their ability to realize the sizable market potential.

Commercial Realities Affect Enterprises' Attitudes toward Investing and the Poor

Cambodian water firms—which invest and operate autonomously on commercial terms—display a strong orientation toward serving the poor. In contrast, few firms in Bangladesh or Benin—where investment is largely or wholly funded by government and donors—consider the poor to be their target market, and many believe that the policies under which they currently operate do not provide the poor with equal access to their service. In all countries, firms believe that costs are beyond the reach of the poor and that no incentives exist to reach these markets.

Sanitation enterprises in all countries recognize that the market for sanitation is growing, but they are concerned about the regularity of demand. Perceptions of the poor as an attractive customer segment vary. In Bangladesh and Indonesia, a majority of enterprises consider the poor as target customers. In contrast, in Tanzania, most do not. More than three-quarters of Bangladeshi enterprises are concerned that the poor do not pay on time, a view shared by smaller majorities in Indonesia and Tanzania. More than three-quarters of enterprises in Tanzania indicate that the poor live in areas that are expensive to service because of transport and infrastructure problems.

Costs and Policy Factors in Water and Commercial Factors in Sanitation Constrain Profitability and the Ability to Offer Value to Poor Customers

In the water sector, more profitable firms tend to want to invest more and to see the poor as a target customer. Profitability is very sensitive to the business model, which is shaped by government policies. In Bangladesh, private entities coinvest in networks with the government and donors. Customers are charged a flat monthly fee, which results in flat revenues in the face of increasing consumption. The tariff structure means that few networks are financially viable. In Benin, the business model is to cover a larger service area through manned standpipes. A top-down investment program designs and builds all networks, which are too large given the scale of the market. The tariff structure is determined by policy-driven financial models that overestimate market sales, leading to high tariffs, which in turn reduce consumption and revenues for the enterprise. In Cambodia, financing, design, construction, operation, and management are wholly private and are well calibrated to market conditions. Nearly all Cambodian enterprises yield positive returns on investment, and revenues enable adequate provisioning for depreciation. Energy is a key issue for water firms in all three countries, accounting for about 40 percent of costs in Bangladesh and Benin and 60 percent in Cambodia.

In sanitation, enterprises find it difficult to be profitable and to offer attractive value propositions to the poor because a fragmented supply chain increases costs for poor customers as well as sanitation suppliers. Sanitation facilities require the aggregation of different component materials and significant coordination of different actors (manufacturers, suppliers, builders, providers of services such as pit emptying). Most of this burden falls on households. The aggregation of materials, many of which have low ratios of value to weight and volume, also carries with it a high embedded cost of transport, which pervades the supply chain from national-level manufacturers and importers down to the local microenterprises with whom households deal. There are no well-resourced players for whom on-site sanitation is a large enough business to warrant intensive efforts to develop and market solutions and coordinate activities across the supply chain. As a result, although poor households aspire to much higher levels of sanitation, they simply make do with what they have rather than purchase options that do not, in their minds, justify the costs.

What Governments Do Matters for Water Enterprises, What Governments Don't Do Matters for Sanitation Enterprises

In addition to market-related risks, water firms face a variety of policy and institutional obstacles, including the bureaucratic hassle of applying for permits and participating in public tenders, the insecurity of licenses, and the lack of effective dispute-resolution mechanisms. In contrast, the impact of policies in the sanitation sector is limited. Enterprises working in the sector would like governments to concentrate on removing risks to entry by providing market intelligence and promoting the entry of enterprises that are able to undertake transformative research and development on new technologies and materials.

Water

Overview of the Water Sector

Throughout the developing world, millions of people lack access to safe water supplies. In the three countries covered in Part 1 (Bangladesh, Benin, and Cambodia), 46 million people lack access to clean water. The problem costs these countries 0.2–0.7 percent of gross domestic product (GDP) a year—at least $275 million in total.

Part 1 examines piped water schemes in rural areas of Bangladesh, Benin, and Cambodia, where the local private sector already plays a major role in the delivery of water (for the purposes of this study, the term *rural* also includes small towns outside of the main urban areas). The majority of households in all three countries get their water from private and communal sources. Little systematic information is available about these markets; most information on the private water sector focuses on large service providers.

Bangladesh, Benin, and Cambodia are countries where the Water and Sanitation Program (WSP)—a multidonor partnership administered by the World Bank to help poor people obtain affordable, safe, and sustainable access to water and sanitation services—is actively supporting client governments in engaging the domestic private sector. The WSP is well placed to offer practical follow-up of the study results in these countries.

The study examines the performance of networks in each country and investigates the preferences of poor households in locations served by them. It also examines commercial and investment climate factors that may affect firms' actual or perceived costs and risks, driving their decisions about increasing investment in their business. Specifically, the study seeks answers to the following questions:

- Is lack of interest by the domestic private sector a rational response to weak market potential, or are lack of firm viability and the use of inappropriate business models preventing it from taking advantage of market opportunities?
- Are policy and investment climate factors increasing the (actual or perceived) cost and risk associated with doing business?

Market Potential for Rural Piped Water Schemes

In the three study countries, the potential market the domestic private sector could be serving is large. By 2025, about 20 million people in Bangladesh, Benin, and Cambodia are projected to get their water from rural piped water schemes—10 times the current number. This market will be worth at least $90 million a year, up from about $23 million in 2012.

Market growth is being driven by a combination of economic and policy factors. Population and income growth are important, but country-specific drivers are at play as well:

- In many locations in Bangladesh, current sources are unsustainable, because of contamination and the growing scarcity of water. A national policy aims to respond to these problems through public/private/community coinvestment in piped water networks.
- In Benin, a key driver is the recent adoption of a policy to contract out management of networks built by the public sector.
- In Cambodia, the costs of alternative sources, the absence of public supply, and a liberal (if somewhat unregulated) government approach to licensing private networks are creating commercial opportunities for autonomous private investment.

Constraints to Serving the Market

A mix of commercial and policy factors is constraining the expansion of private schemes. The commercial factors are broadly similar across the three case study countries. The policy factors are more country specific. Although all three countries recognize the role of the private sector in increasing access and improving quality of service, each has policies that make it difficult for private firms to be profitable, thereby dampening firms' interest in investing.

Weak Demand
Households, especially poor households, purchase too little water from networks for operators to achieve optimal capacity utilization or to warrant significant investments in additional capacity. Poor households need higher volumes of water, but their purchases are limited by cost and their assessment of the value of network water with respect to alternatives.

Tariffs and connection fees are too high for many poor rural households. Standard daily per capita consumption of 40 liters of water would cost them 2.4 percent of income in Bangladesh, 5.6 percent in Benin, and 4.1 percent in Cambodia.[1] Although tariffs in Bangladesh and Cambodia do not yet breach the traditional ceiling of 5 percent of income, costs may be running up against budget constraints of poor households and what they are prepared to pay. In Bangladesh, for example, where expenditure on water by poor households consumes less than 1 percent of total household expenditure, a majority of households indicated that they could afford to pay only about half the current tariffs. For many poor households, the cost of a private connection is an ever greater barrier to use of network water, with costs averaging 27 percent of monthly incomes in Bangladesh, 116 percent in Benin, and 34 percent in Cambodia.

Most households have access to inexpensive alternative sources of water (if only for parts of the year), including rainwater, wells, springs, and boreholes. They are savvy about making trade-offs between price and value in choosing their water source. In the short run, competition from other sources will limit demand for piped water. In the longer run, the availability and opportunity cost of alternatives will likely shift incentives in favor of networks, especially if operators can ensure consumers of the quality of the service they offer.

Lack of Firm Viability and Inappropriateness of Business Models

At a certain network size, piped water systems offer considerable economies of scale in providing potable water. But reaping these economies requires operating above certain minimum levels of sales, and economic and financial sustainability requires charging prices that cover all costs. Getting this balance right is challenging.

Different business models have emerged in the three countries as a result of market and policy drivers. Each has achieved a different degree of success.

In Bangladesh, private sponsors coinvest in networks with the government and donors in localities where groundwater cannot be safely used. Customers are served through private connections. They pay a flat monthly fee, which results in low revenues despite high volumes. Combined with the fact that most networks have too few connections given the investment cost to households, the tariff structure means that few networks are financially viable.

In Benin, the business model is to cover a larger service area through manned standpipes. A top-down investment program designs and builds all networks, which are too large given the scale of the market. The tariff structure is determined by policy-driven financial models that grossly overestimate market sales, leading to very high fees and tariffs. Tariffs provide a large profit margin for every unit of water sold and most operators therefore make a profit on their leases, but they keep consumptions levels low. As a result, aggregate revenues do not cover investment costs.

In Cambodia, financing, design, construction, operation, and management are wholly private. Networks serve households through metered connections.

Nearly all networks yield positive returns on investment, and revenues enable adequate provisioning for depreciation. Designed capacity is well calibrated to the market and continuity of service is good. But lack of access to water sector expertise may be leading to suboptimal choices of design and equipment, and the potability of water may not be ensured.

Attitudes toward Investment and Serving the Poor

Water firms in Bangladesh and Benin, where the public sector and donors largely determine which assets are built, are circumspect in their attitudes toward investment. Few firms in Bangladesh were planning investment, and the investment that was planned focused on expanding the coverage of or repairing existing networks. In Benin, nearly half of firms interviewed were planning investment, but spending seemed to be going toward maintenance to allow assets to continue functioning. In contrast, in Cambodia, three-quarters of enterprises interviewed were contemplating investments in existing networks, with a strong emphasis on network and water production expansion, and half of the enterprises were interested in investing in new sites.

Enterprises identified a range of market-related risks that affect their investment plans. In Bangladesh, the main concern was that costs make profitability uncertain for both existing and new networks. This concern reflects current conditions in the market, where most firms are not profitable. In Benin, enterprises cited their lack of experience in developing (as opposed to operating) systems. They cited a wide range of risks, including concerns about water availability, lack of sufficient demand, and the high cost of investment and expressed uncertainty about which investments to make. In Cambodia, firms' greatest concern was access to finance.

Cambodian firms display a strong orientation toward serving the poor. In contrast, few firms in Bangladesh or Benin considered the poor as their target market, and many believed that their policies did not provide the poor with equal access to their services. In all countries, firms believed that costs are beyond the reach of the poor and that no incentives exist to reach these more difficult markets.

Unsupportive Investment Climate

In addition to market-related risks, firms face a variety of policy and institutional obstacles. In Bangladesh, the pricing and ownership structures do not seem to allow networks to recover—or even earn a return on—their capital costs, and investment is contingent on government or donor cofinancing. As a consequence, private operators appear reluctant to expand networks or sponsor additional networks. In Benin, the main barrier to expansion is the lack of capacity of the public sector in designing appropriately scaled networks and tendering them for private operation and the nature of the leases under which firms operate networks. In Cambodia, the incomplete nature of the legal framework on urban and semi-urban water supply and lack of clarity and consistency about the rules governing private investment in water networks may be constraining the types of investment that private firms are prepared to make.

To manage the uncertainty associated with licenses, enterprises usually consider building networks only if they already own suitable land on which to build the system infrastructure.

The lack of good physical and financial infrastructure also stifles investment. Firms in all three countries singled out unreliable power supply as a key constraint to doing business. Energy accounts for 39 percent of operating costs in Bangladesh and Benin and 65 percent in Cambodia. Where networks use diesel fuel (to generate electricity full time or as a backup or to run intake pumps), energy costs are significantly higher.

The limited reach of the financial sector and the costs of accessing finance also limit firms' ability to invest. In Cambodia, for example, all loans must be collateralized by real estate.

Recommendations

The recommendations of this report are intended to help policy makers remove or relax the main constraints preventing the private sector from providing piped water to the poor. Although they are based on the case studies, they are relevant for other countries as well.

Stimulate Demand by the Poor

1. Improve affordability by right sizing: design and build assets that are appropriate for small-scale networks, so that cost-recovery prices can be kept as low as possible.
   • Realistically assess demand and adopt design and construction standards and procurement rules to align network design with it.
   • Modify tendering systems to identify inputs in terms of performance and quality standards rather than by specifying particular brands or suppliers.
2. Improve affordability by smoothing and subsidizing expenditures: experiment with initiatives that enable poor rural households with volatile cash incomes to spread connection payments (and perhaps usage charges) over time.
   • Where facilities for cash transfers to the poor already exist, consider providing targeted demand-side support for the extreme poor.
   • Where networks are leased to private operators or involve coinvestment by government or donors, consider including a requirement in lease contracts or project designs that concessional terms for connections be offered to poor households. Where network construction and operation are completely independent of government and donors, consider delivering support directly to households, rather than trying to impose community service obligations on operators.
   • Develop financing schemes that enable operators to offer customers installment plans for paying for private connections.
3. Establish appropriate standards to help firms signal water and service quality to the market.

- Identify service and quality standards and means of achieving them that are both consistent with regulatory capacity and simple enough for consumers to understand.
- Help firms implement standard procedures for ensuring water quality and targeting information campaigns to their customers.

Improve Firm Viability and Business Models

4. Improve profitability by removing impediments to efficient pricing, without which private operators cannot be financially viable.
   - Introduce metering, so that firms are paid for increased usage (Bangladesh).
   - Where tariffs and charges are regulated, recalibrate models to avoid setting tariffs so high that they restrict consumption excessively (Benin).
5. Improve profitability by optimizing the operation of the network under contract, where contracted-out networks face competition from other publicly owned water sources.
   - Assess the feasibility of regulating exclusivity and alternative delivery in network locations (by including public water points in operator contracts with appropriate pricing, for example).
   - Develop regulated arrangements for sharing connections or resale of water from private connections to increase consumption and capacity utilization.
6. Expand private connections by establishing incentives for incremental upgrades of existing networks to offer more private connections, which provide the convenience that consumers strongly value.
   - Grant concession contracts or enhanced lease contracts in which the private operator implements publicly funded investment in network expansion/densification (Benin).
   - Improve the planning, marketing, and design of networks to locate water points where households need them, and promote the use of private connections (Bangladesh and Benin).
7. Improve supply chains and technical support by improving professional capabilities for the design, construction, and maintenance of small-scale piped water networks.
   - Foster the creation of professional associations to train and provide accreditation for consultants who design networks or provide other expertise to small-scale water operators.
   - Support business brokering initiatives that could work with financial institutions to assess the risk and feasibility of network investments by small enterprises.
   - Reduce the size of lots in the public procurement of water infrastructure development, in order to allow local players to compete and build capacity.

Improve the Investment Climate and Sectoral Policies

8. Provide market intelligence to improve information for potential investors about investment opportunities, so that enterprises are aware of the

availability of water resources and market potential in areas outside of their current areas of operation.

- Improve sector investment planning to identify—and publicize—markets with potential for private participation.
- Provide technical support to local authorities to develop projects that can be taken to market.

9. Increase access to finance to address the low level of financial inclusion and the limited level of financing for small water projects.
   - Develop financing facilities to support cash flow–based financing for water projects, including the use of blended funds, credit enhancements, guarantees, and cost-sharing arrangements, and provide appropriate project development and appraisal support to financial institutions.
   - Develop robust loan documentation that is consistent with national legal frameworks, and assist with legal reform and clarification to facilitate market-based financing of and investment in water projects.

10. Increase access to land and energy, by facilitating land access for private water schemes and addressing the high cost and limited and unreliable supply of energy.
    - Where concession law structures are in place, use them to bring small-scale water projects to the market with provisions for land access and infrastructure development (Cambodia).
    - Consider offering incentives for generating power for water projects in locations that are poorly served by the grid.

11. Improve government policy and practice by improving policy clarity and functionality to facilitate the provision of piped water in more marginal locations.
    - Prepare operational guidance on the role of the private sector, and move from project- to policy-based approaches to increase transparency and competition and avoid distortions created by inconsistency and idiosyncratic subsidization (Bangladesh, Benin, Cambodia).
    - Improve arrangements for determining tariffs, and introduce incentives for expanding coverage and meeting service standards (Benin).
    - Where the prevailing model is public-private partnerships, improve incentives for sustainable service delivery by including incentives to expand coverage and meet service standards; improving arrangements for determining fees paid by network operators; tying them to likely revenues and costs; and clarifying responsibilities for repair, replacement, and expansion of the network (Bangladesh, Benin).
    - To encourage supply in hard-to-reach or less profitable locations, where the prevailing model is autonomous private investment, develop a system of competitive tendering of rights to localities using a more traditional public-private partnership model, and ensure that interventions that stimulate private provision create a level playing field (Cambodia).

12. Improve government policy and practice by strengthening dispute-resolution arrangements, the absence of which deters investment.

Tapping the Markets • http://dx.doi.org/10.1596/978-1-4648-0134-1

- Provide training programs for public and private parties to contracts to improve their understanding of obligations, and introduce mechanisms to support regular business planning and performance review processes as a companion to dispute-resolution arrangements.
- Empanel independent reviewers and auditors to help contracting parties resolve disputes.

Note

1. Per capita consumption of 40 liters per day is the World Health Organization standard. For many water users, actual consumption is much lower.

CHAPTER 1

What Is the Problem?

Throughout the developing world, about 780 million people lack access to safe water. Outside of large urban centers, private providers dominate the water sector. In Bangladesh, for example, where access to safe water is relatively good, 28 million poor people and another 11 million people living above the poverty line still lack access to improved water sources.

In the three case studies examined in Part 1—Bangladesh, Benin, and Cambodia—only about 11 percent of the population gets its water from state utilities; the rest rely on a combination of self-supply, private provision, and community-run systems. In many cases, use of these alternative sources of water endangers health and well-being and reduces productivity.

Access Is Inadequate

Inadequate access to safe water is costly. According to conservative estimates by the World Health Organization, in Bangladesh, Benin, and Cambodia alone it costs about $275 million a year, 0.2–0.7 percent of each country's gross domestic product (GDP) (WHO 2012; table 1.1).

Access to improved water increased over the past two decades, but progress has varied. In Bangladesh, overall access has been high since the late 1990s, but progress toward increasing coverage of the 25 percent of the population that was without improved water in 1990 has been slow. In Benin and Cambodia, changes have been more pronounced, albeit from much lower starting points (figure 1.1). None of the three countries is close to universal access, even in urban areas.

Improved water supplies can come from a range of sources, including protected wells, springs, and boreholes. Most people in developing countries rely on these sources. Piped water systems have the potential to offer economies of scale and improved convenience. They are increasingly being installed in places where the population is sufficiently dense or alternatives are not available (for example, where local groundwater is contaminated or costly to access).

Table 1.1 Estimated Annual Economic Costs of Inadequate Water Supply in Bangladesh, Benin, and Cambodia, 2012

Country	Number of people without improved water (millions)	Cost (millions of dollars, in 2010)	Share of GDP (percent)
Bangladesh	39.0	176	0.2
Benin	2.3	49	0.7
Cambodia	5.0	51	0.5

Sources: Country case studies; WHO 2012. Unless otherwise indicated, data for tables and figures have been taken from the country case studies (see references).

Note: Cost figures show economic benefits forgone as a result of not achieving universal access to improved water. They include the forgone health and time-saving benefits of access to improved water supplies. GDP = gross domestic product.

Figure 1.1 Access to Improved Water in Bangladesh, Benin, and Cambodia, 1990–2010

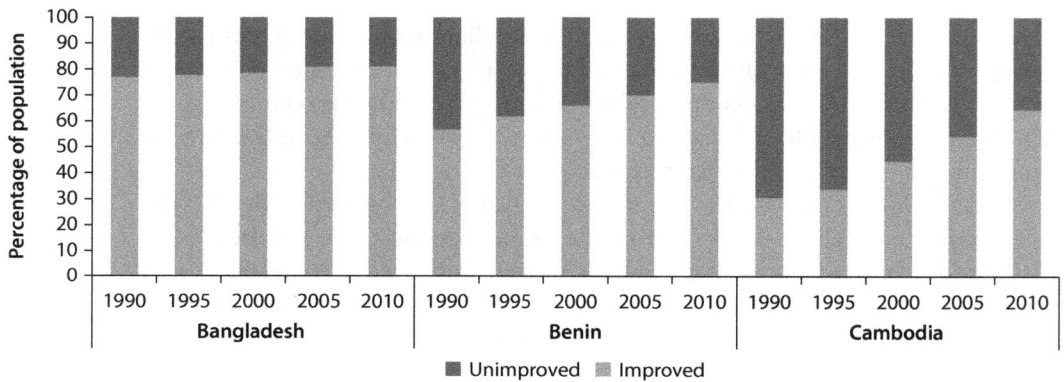

Sources: WHO and UNICEF 2012a, 2012b, 2012c.

The Costs Are Borne Largely by the Poor

In developing countries, the direct costs of inadequate access to safe and convenient sources of water are borne predominantly by the poor. Poor people are much less likely to be served by public utilities, and they are less equipped to deal with the consequences of using unsafe water. Loss of income as a result of waterborne illness can have a catastrophic effect on poor households, which are often unable to afford treatment or to survive long without income.

Governments Cannot Solve the Problem

Private firms may be underinvesting in the water sector because people may not be willing or able to pay prices that reflect all of the social benefits of using clean water. The existence of these "externalities" does not necessarily provide a rationale for government provision of water, however. Moreover, even if it did, in most developing countries with large numbers of poor people, the government lacks the financial and organizational capacity to meet the need for improved water supplies from public resources.

In the countries covered by this study, most poor (and many nonpoor) house-holds look to the private sector to help meet their water needs. It is the private sector that builds or supplies materials and components for self-supply, sells water from standpipes or water trucks, and builds and operates piped water systems.

References

Country Studies

DevCon (DevConsultants Limited). 2013. *Water Supply Bangladesh: Global Study for the Expansion of Domestic Private Sector Participation in the Water and Sanitation Market.* Dhaka: DevCon.

GRET (Groupe de Recherché et d'Echanges Technologiques). 2013. *Final Report Cambodia: Global Study for the Expansion of Domestic Private Sector Participation in the Water and Sanitation Market.* Phnom Penh: GRET.

Hydroconseil. 2013. *Benin: Deep Dive Analysis Report. Global Study for the Expansion of Domestic Private Sector Participation in the Water and Sanitation Market.* Cotonou: Hydroconseil.

Other References

WHO (World Health Organization). 2012. *Global Costs and Benefits of Drinking-Water Supply and Sanitation Interventions to Reach the MDG Target and Universal Coverage.* WHO/HSE/WSH/12.01. Geneva, Switzerland: WHO.

WHO (World Health Organization)/UNICEF (United Nations Children's Fund) Joint Monitoring Program. 2012a. "Estimates for the Use of Improved Drinking Water Sources, Bangladesh." http://www.wssinfo.org/fileadmin/user_upload/resources/BGD .xlsm.

———. 2012b. "Estimates for the Use of Improved Drinking Water Sources, Benin." http://www.wssinfo.org/fileadmin/user_upload/resources/BEN.xlsm.

———. 2012c. "Estimates for the Use of Improved Drinking Water Sources, Cambodia." http://www.wssinfo.org/fileadmin/user_upload/resources/KHM.xlsm.

CHAPTER 2

Why This Study?

A vibrant and diverse local private sector is critical to the delivery of services, as a large body of research conducted over the past decade shows.[1] More and more governments have been emphasizing the role of the domestic private sector in their national strategies, and domestic players have been moving from niche provisioning to mainstream operations. Multiple levels of capacity constraints prevent the domestic private sector from delivering at scale, however, and few lessons are available to help them deal with these constraints.

This study examines the involvement of the domestic private sector in the construction and operation of rural piped water networks. Its aim is to understand the extent to which private sector schemes can provide the poor with safe water.

The study considers two sets of factors—commercial factors and investment climate factors—that affect firms' actual or perceived costs and risks and, in turn, their decisions about investing in water networks (figure 2.1). It examines both sets of factors by seeking answers to the following questions:

- Is lack of interest by the domestic private sector a rational response to weak market potential, or are lack of firm viability and the use of inappropriate business models preventing it from taking advantage of market opportunities?
- Are investment climate factors increasing the (actual or perceived) costs and risks associated with doing business?

To shed light on these issues, the study team conducted research into the water sector and its policy environment, surveyed operators of water networks, held focus group discussions with water users, and interviewed other stakeholders, including government officials, in Bangladesh, Benin, and Cambodia. The country studies examined "rural growth settlements"—villages with fewer than 10,000 households (sufficient density to warrant a network solution); some infrastructure (roads, electricity, telecommunication coverage, education and health services); and economic dynamism despite reliance on rural practices and livelihoods.

Figure 2.1 Study Analytical Framework

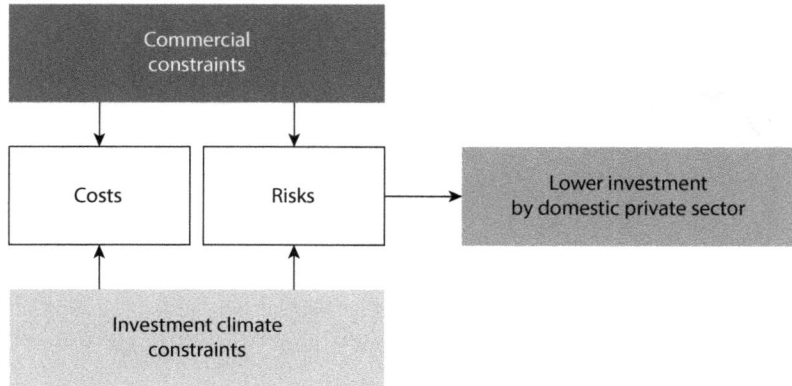

Each study involved the preparation of a country analysis that examined the market structure, relevant supply chain, and policy environment; a survey of firms that deliver services to poor households; and focus group discussions with poor water users. Across the three countries, 89 firms were surveyed, and focus group discussions were conducted with 1,100 people.

Note

1. For evidence, see Kariuki and Schwartz 2005; Triche, Requena, and Kariuki 2006; Valfrey-Visser and others 2006.

References

Kariuki, Mukami, and Jordan Schwartz. 2005. *Small-Scale Private Service Providers of Water Supply and Electricity: A Review of Incidence, Structure, Pricing and Operating Characteristics.* Washington, DC: World Bank.

Triche, Thelma, Sixto Requena, and Mukami Kariuki. 2006. *Engaging Local Private Operators in Water Supply and Sanitation Services. Initial Lessons from Emerging Experience in Cambodia, Colombia, Paraguay, Philippines and Uganda, vol. 1. Overview of Experience.* Water Supply & Sanitation Working Note 12. Washington, DC: World Bank.

Valfrey-Visser, Bruno, David Schaub-Jones, Bernard Collignon, and Emmanuel Chaponnière. 2006. *Access through Innovation: Expanding Water Service Delivery through Independent Network Providers: Considerations for Practitioners and Policymakers.* London: Business Partners for Development.

CHAPTER 3

Water Networks and the Role of the Government

Rural piped water schemes in Bangladesh, Benin, and Cambodia are similar, but the prevailing models in each country are very different. These differences largely reflect the role of the government in the sector and the impact it has on opportunities for private and nonstate actors to invest in or operate piped water networks.

Salient Features of Networks

Table 3.1 summarizes the salient features of rural piped water systems in the three countries studied. In Bangladesh, most networks use groundwater. Networks typically have treatment plants only where surface water is used or systems are built to address water contamination. All networks are connected to the electricity grid and use electric pumps. Most connections are made directly to households or institutions. No connections are metered.

Network structures in Benin are designed to standard specifications set by the government. They are therefore more uniform than networks in Bangladesh or Cambodia. All networks use groundwater; they do not have separate treatment plants. Only 25 percent of networks use electricity from the grid; of these, 40 percent use a backup generator. The majority use diesel power to operate pumps or generate electricity for production and distribution. Water is sold primarily through operator-managed standpipes, which are designed to serve 250 people. All connections are metered.

In Cambodia, schemes range widely in size, but about three-quarters of them draw water from rivers or ponds, using a pump to bring water into storage. Sixty percent rely on diesel or petrol for pumping and generating electricity. Seventy percent have a water treatment plant and combine an underground water storage tank with a water tower. Nearly all connections are made directly to homes and metered. Very few connections are for commercial or institutional customers.

Table 3.1 Salient Features of Rural Piped Water Systems in Bangladesh, Benin, and Cambodia, 2012

Feature	Bangladesh	Benin	Cambodia
Investment	Majority public financing (government or development partner); private firms and NGOs provide 20–30 percent of investment	Fully public financing	Fully private financing
Operator	Mostly NGO and community organizations, some private firms	Private firms	Private firms
Systems	Deep well water, mostly private connections; production metering only	Deep well water, mostly standpipes; production and consumption metering	Surface water, mostly private connections; production and consumption metering
Tariff structure	Flat fees	Volumetric	Volumetric

Note: NGO = nongovernmental organization.

Table 3.2 Investment in Construction of Water Systems in Bangladesh, by Sponsor, 2012
Percentage of average investment cost

	Percentage of average investment cost contribution		
Constructed by	Government and donor	Private entity (firm or NGO)	Community
Private sponsors	70[a]	22	8
Government sponsor, no donor funding	100	0	0
Government sponsor with donor funding	92	0	8
NGO sponsors	91	0	9

Source: DevCon 2013.
Note: NGO = nongovernmental organization.
a. Initially, two sponsors contributed 40 percent or more of the investment cost, but the project rules were later changed, reducing the required contribution to 30 percent.

Role of the Public and Private Sectors

At one end of the spectrum is Benin, where the public sector finances, designs, and constructs water systems and only the operation and management of systems is delegated to private firms. At the other end of the spectrum is Cambodia, where the private sector finances, designs, constructs, operates, and manages all water systems. In Bangladesh, private operators have some equity stake in the networks and are involved in their design and construction (table 3.2).

Reference

DevCon (DevConsultants Limited). 2013. *Water Supply Bangladesh: Global Study for the Expansion of Domestic Private Sector Participation in the Water and Sanitation Market.* Dhaka, Bangladesh: DevCon.

CHAPTER 4

Is Market Potential Sufficient to Justify Private Investment?

The water market is large: a conservative estimate of household outlays on water suggests that people in rural and semi-urban areas of Bangladesh, Benin, and Cambodia spend at least $620 million a year on water. Poor people alone spend more than $270 million a year on water (table 4.1). These estimates exclude households connected to public utilities (which operate primarily in metropolitan areas) and consumption of water from free-of-charge sources. They are thus a measure of what people are paying for water from small-scale operations.

The majority of people (90 percent in Bangladesh, 67 percent in Benin, and 92 percent in Cambodia) get their water from private (including self-supply) and communal sources. In addition to operating piped water systems, private operators supply rural households with facilities for self-provision or sell bottled or carted water. In Bangladesh, the private sector has helped more than 60 million rural residents meet their own water needs using hand pump tubewells.

The importance of small-scale rural schemes varies across the three countries. Only about 0.1 percent of the population of Bangladesh gets its water from such schemes (table 4.2). In contrast, 7.2 percent of the population in Cambodia and 15.9 percent in Benin use these schemes.

The contribution of small-scale piped water schemes is projected to grow in all three countries, with annual water sales expected to increase from $23 million in 2012 to at least $90 million by 2025.

A combination of market and nonmarket drivers is creating opportunities for private enterprises (table 4.3):

- In Bangladesh, private hand pumps are a cheap, widespread source of safe water for most of the rural and semi-urban population. The main driver of market growth for piped water services is the contamination of shallow water (primarily from arsenic, iron, and salt). The government's 2011–25 Sector

Table 4.1 Estimated Size of Water Market in Bangladesh, Benin, and Cambodia, 2012

	Bangladesh	Benin	Cambodia	Total
Size of market (millions of people not connected to state water utilities)				
Entire country	135.6	6.1	13.0	154.7
Rural	107.2	5.3	11.3	123.8
Poor	45.4	3.5	4.1	53.0
Value of market (millions of dollars)				
Entire country	512.7	74.1	175.9	762.8
Rural	405.4	64.4	152.9	622.7
Poor	171.7	43.1	55.5	270.2

Note: Market values were estimated by assuming that average outlays on piped water in the households covered by the country studies are typical for all households in each country. These outlays do not typically cover all water consumed by households, and prices paid and costs incurred vary across and within each country.

Table 4.2 Rural Piped Water Schemes under Private Management in Bangladesh, Benin, and Cambodia, 2012

Item	Bangladesh	Benin	Cambodia	Total
Number of rural piped schemes	75[a]	350	300[c]	725
Percent privately managed	100[b]	35	100	68
Thousands of people served	200	500	1,030	1,730
Percentage of population served	0.1	15.9	7.2	0.9

a. 150 schemes were found, but only half were operational.
b. Includes operation by nongovernmental and community organizations operating on a commercial basis.
c. Some sources in the country study cited estimates as high as 800 (GRET 2013).

Development Plan for water estimates that by 2025, 10 percent of the rural population (about 11 million people) will be served by rural piped water schemes. (In 2012, about 200,000 people were served by these schemes.) Meeting this goal will require investment of about $800 million.
- In Benin, annual sales of water from privately or community-managed networks in small settlements could reach $22 million by 2025.
- In Cambodia, where private investors are building networks where public utilities are not operating, sites with suitable characteristics could support a doubling of the number of people supplied by private networks, to about 2 million people.

In each country, opportunities for private participation in small towns and rural areas are shaped by the extent of national guidance (policy clarity), the government's willingness to engage private partners, and its ability to implement policy effectively (operationalization) (table 4.4). The processes of political and national decentralization that have taken place in all three countries in recent years are also shaping the landscape: in all three countries the "political distance" between network operators, customers, and the local political personalities that oversee their performance is very small.

In Benin, opportunities for providing rural piped water depend on the space created by the government and its capacity to invest in networks that private

Table 4.3 Drivers of Market Opportunities in the Water Sector in Bangladesh, Benin, and Cambodia

Country	Immediate driver	Opportunity created
Bangladesh	• Strong economic growth and social development: per capita GDP grew from $1,164 (purchasing power parity) in 2005 to $1,643 in 2010, and poverty rates dropped from 40 percent to 31 percent • Stable and relatively large rural population (about 100 million) until 2030 • Unsuitability of current groundwater point sources, as a result of contamination (from arsenic, iron, and salt); scarcity of groundwater in northeast areas and switch to surface water sources; lowering of groundwater levels as a result of excessive abstraction from irrigation wells) • Large sector investment program budget (twice traditional funding levels from the public sector)	• Demand for higher levels of service as a result of move away from hand pump tubewells • Need for higher-investment solutions, which require more centralized treatment, storage, and transport • Increased recognition by government of need to leverage private investments; greater openness to develop conducive business environment for private sector participation in water
Benin	• Increase in construction of networks under public investment program: annual rural budget tripled between 2001 and 2010 (to $26 million); target of 112 schemes by 2016 • Decision by commune (local government) authorities to put management of network out to tender • New strategic framework for developing the private sector, which specifically includes the water sector	• Increased private sector lease contracts • Smaller lot sizes of procurement introduced, allowing small firms to bid for projects
Cambodia	• Strong economic growth and economic stability (average GDP growth of 8 percent between 2005 and 2010) • Large volume but low quality of water (most water comes from the Mekong and Tonle Sap rivers) • Inadequate public supply of water	• Support to increase household ability and willingness to pay • Relatively easy extraction and value-adding through treatment • Secure niche markets

Note: GDP = gross domestic product.

Table 4.4 Clarity and Operationalization of Government Water Policy in Bangladesh, Benin, and Cambodia, 2012

Item	Bangladesh	Benin	Cambodia
Policy clarity			
Responsibility for service delivery (national or local)	National but in transition. Ministry of Local Government, Rural Development and Cooperatives has statutory mandate. In rural areas, mandate is implemented by Department of Public Health and Engineering (DPHE), a line agency, with support from local administrative institutions. Elected local governments operate at district and union but not subdistrict levels. Recent projects emphasize role of local governments as eventual asset owners.	Clear allocation of responsibility to local authorities under Law N°97-029 of January 15, 1999, pertaining to organization of communes.	Less clear than in other countries and in transition. Memorandum of Understanding between Ministry of Rural Development and Ministry of Industry, Mining and Energy (MIME) (traditionally responsible for urban development) allocates responsibility for supervision of rural private piped water to MIME. MIME also directly delivers services through state agencies in provincial capitals. Recent decentralization laws and strategies place public service delivery at provincial, district, and commune levels.
Private sector engagement	Unclear. Although policies and strategies promote private sector development, there are no guidelines on operational definition or classification of private sector, scope of participation, or financing mechanisms. Different models emerge through projects.	Directional. National strategy stipulates that network operations need to be managed by private sector, and guidance is available for selecting them through a tender process under four possible arrangements, including possibility of engaging users associations or private firms.	Laissez-faire. A concession law exists, but most water businesses operate under a licensing regime that is not formalized in law and is governed by few regulatory constraints.
Operationalization of policy mandate for private participation			
Market identification and scope determination	Combination. Government sets parameters on location through priority development criteria/list; NGOs and private firms present proposals for partnerships.	Public only. Identification occurs through national development program priorities and decision of local communes.	Mostly private. Except in specific development projects, private sector selects area in which it operates and scale of operations.

table continues next page

Table 4.4 Clarity and Operationalization of Government Water Policy in Bangladesh, Benin, and Cambodia, 2012 *(continued)*

Item	Bangladesh	Benin	Cambodia
Investment financing and design	Combination. Water is funded mainly through national development programs under a cost-sharing arrangement. DPHE provides guidance on design. Generally, the private party (directly or through others) designs and builds the system, with supervision support/ approval by DPHE.	Public only. Infrastructure is developed through public engineering, procurement, and construction under specific guidelines.	Entirely private.
Fees, tariffs, and revenue generation	Negotiated. Tariffs are negotiated with communities and local governments. Firms defray administrative costs of supervision (for example, meeting costs) by local institutions. Level of return is not set.	Directional. Tariff formulas are used to determine the lease fees and level of tariffs. Level of margin is not set.	Market based. Main restriction on tariffs is what the market will bear (there is stiff competition from alternative sources).
Performance accountability	Vague. In most cases, local associations are formed to provide customers with voice. Minimal national support through DPHE.	Systematic, but not applied. Formal accountability process is established in the contract, but does not work consistently in practice. No formal process exists for identifying well- vs. poorly performing firms, although the sector is small and word is likely to get around. No national-level support for contract and conflict management visible.	Self-imposed. Licenses have minimal service-level requirements, which are not systematically monitored. As firm investments are high and firms rely on the business, performance is closely linked to firm viability.

Note: NGO = nongovernmental organization.

operators can run. In Cambodia, opportunities arise from the inability of public services to expand into markets with potential for profitability and the extent to which the government allows private initiatives to function without constraint.

Reference

GRET (Groupe de Recherché et d'Echanges Technologiques). 2013. *Final Report Cambodia: Global Study for the Expansion of Domestic Private Sector Participation in the Water and Sanitation Market.* Phnom Penh: GRET.

CHAPTER 5

What Affects Demand for Water?

Aggregate market potential for piped water is significant in the three countries studied. But demand by the poor is weak, because poor households are willing to allocate their limited cash only to a compelling service offer, especially when they can use alternative sources of water at very low monetary cost.

Focus group discussions with households in communities served by piped water networks explored a range of issues concerning demand for water, including desired characteristics of water supply, willingness and ability to pay, and perceptions regarding the performance of the network. Table 5.1 describes the sample of respondents.

Cost of Water (Tariffs and Connection Fees)

Tariffs and connection fees are too high for many poor households to afford. If met by water from networks, the standard daily per capita consumption of 40 liters of water would cost poor households 2.4 percent of their income in Bangladesh, 5.6 percent in Benin, and 4.1 percent in Cambodia (table 5.2).[1] Although tariffs in Bangladesh and Cambodia do not breach the 5 percent of income ceiling often applied in the water and sanitation sector, costs may be running up against household budget constraints and the limit to what poor people are prepared to pay.

In Bangladesh, water accounted for less than 1 percent of total household expenditures among the poor, but 53 percent of households indicated that the most they could afford to pay for piped water was $0.78 per month—about half the average actual tariff of $1.50. In Cambodia, poor people living in a water network area indicated that they could pay no more than $0.49 per cubic meter—about 20–30 percent less than the actual tariff of $0.61; people living outside the network indicated that they could afford to pay just $0.45.

Most poor people in Bangladesh and Cambodia have private water connections. In contrast, just 6 percent of focus group participants in Benin had private connections; almost all respondents there indicated that they could not afford a connection at current charges.

Table 5.1 Characteristics of Focus Group Participants
Percentage of households interviewed, except where otherwise indicated

Characteristic	Bangladesh	Benin	Cambodia
Number of villages covered	32	31	27
Number of people interviewed	221 women, 223 men	213 women, 189 men	99 women, 121 men
Classified as poor (percent for locality where focus group discussions held, all countries)	59	39	35
Has very poor quality housing	48	22	21
Lacks own latrine	10	74	—
Owns mobile phone	96	68	88
Average monthly spending on water (dollars)	1.30	8.00	3.80

Note: Poverty measured using national poverty line. — = not available.

Table 5.2 Cost of Water Service as Portion of Household Income in Bangladesh, Benin, and Cambodia, 2012

Country	Monthly income of poor households (dollars)	Water service item	Cost per household (dollars)			Share of household income (percent)		
			Low	High	Average	Low	High	Average
Bangladesh	63	Connection fee	0	50.0	17.0	0	80.0	27.2
		Monthly charge (flat)	0	2.5	1.5	0	4.0	2.4
Benin	173	Connection fee	160.0	729.0	200.0	92.5	421.4	115.6
		Cost per cubic meter	0.6	1.2	1.0	—	—	—
		Monthly cost (at 40 liters per capita per day)	5.7	11.8	9.7	3.3	6.8	5.6
Cambodia	101	Connection fee	20.0	50.0	34.0	19.9	49.7	33.8
		Cost per cubic meter	0.4	1.0	0.6	—	—	—
		Monthly cost (at 40 liters per capita per day)	2.8	6.6	4.1	2.8	6.6	4.1

Note: — = not available.

In Bangladesh, 100 percent of people who were not connected to the water network identified the high connection fee as the reason for not connecting, even though 70 percent of surveyed networks charge no connection fee for shared connections and about 20 percent do not charge for private connections. Other reasons included inability to pay the monthly tariffs (86 percent), lack of availability of connections (46 percent), and poor system performance (45 percent). When asked what they might be willing to pay for connections, 84 percent of respondents indicated a connection fee of less than $12, about 30 percent less than the average actual fee of $17.

In Cambodia, people who were not connected to the water network in their area indicated that a reasonable price to pay for a connection was $30, about 13 percent less than the average actual fee of $34; people living outside the network area were willing to pay an average fee of $26. These costs were lower than but not very far from actual costs.

Table 5.3 Selected Annual Household Expenditures by Poor Households in Bangladesh, Benin, and Cambodia, 2012

Dollars

Item	Bangladesh	Benin	Cambodia
Mobile phone	68	120	—
Housing	51	36	—
Water	16	96	46
Electricity or gas	36	96	—
Memo item			
Average cost of private connection[a]	17	200	34

Note: — = not available.

a. The connection cost covers only the cost of making the physical connection to the network and installing meters (where required).

Another perspective on affordability of connections comes from what people spend on water compared with other items of household consumption. Households in the survey areas pay much more for mobile phones than they pay for water, suggesting that they could afford to connect to a piped water system (table 5.3).

A key issue affecting households' ability to pay for water connections is the variability and lack of regularity of their incomes: in all three countries, income flows among poor rural households are seasonal, and levels of uncertainty about their incomes are high for many households. None of the network operators covered by the studies offered installment payment options, and no credit facilities were available to help potential users cover their water costs. Majorities of focus group participants in all three countries expressed interest in installment payment options.

Competition from Other Sources of Water

Private operators of piped networks have to compete with other sources, including traditional wells, tubewells with hand pumps, watercourses, and rain. The existence of these alternatives reduces firms' revenue potential.

In Bangladesh, where piped water schemes are found largely in areas where contamination has made water from traditional hand pump tubewells unsafe to drink, almost half of households (47 percent) continued to use traditional sources as a complement to piped water (figure 5.1).

Focus group discussions in Benin revealed that the rural poor use the least expensive source of water they can as long as they deem its quality acceptable for the use to which they put it (table 5.4). In areas served by a water supply network, 92 percent of focus group participants obtained some of their water from the network. However, the general pattern is for them to combine piped water with less expensive sources of water, depending on availability and use. People interviewed in Benin, for example, realized that river and pond water is

Figure 5.1 Sources of Water among People with Access to a Water Network in Bangladesh and Benin, by Use, 2012

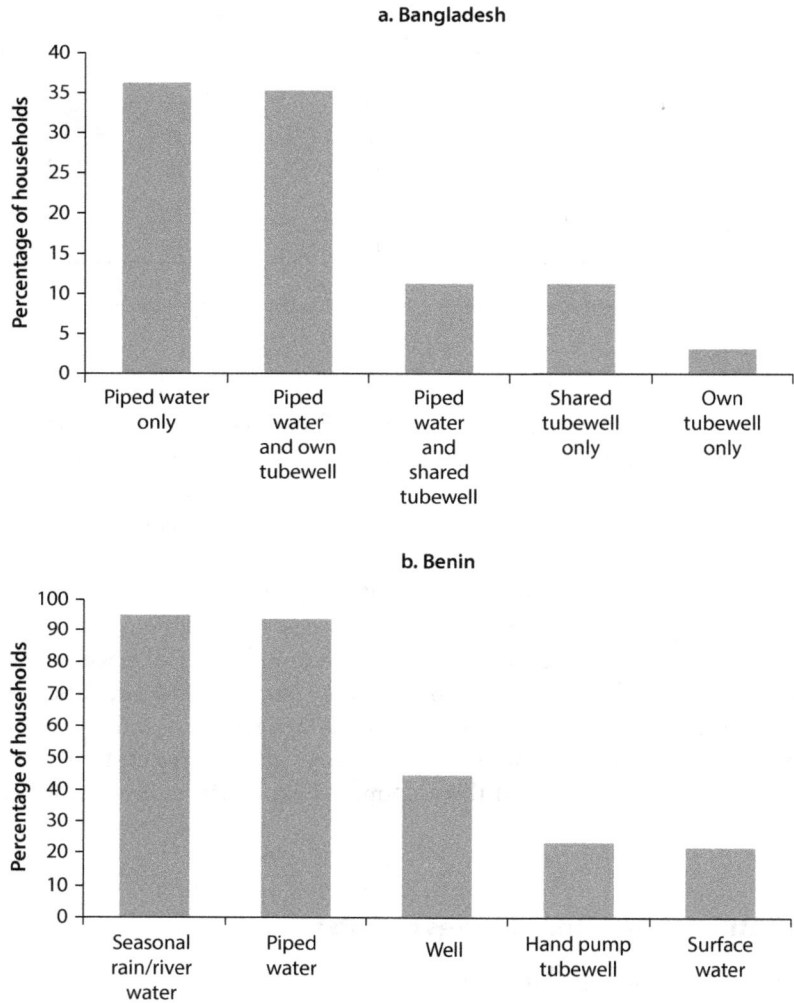

a. Bangladesh

b. Benin

Note: Some households use more than one source, so the total percentage across all sources equals more than 100 percent.

not potable. Despite this knowledge, they often used such water for drinking and cooking (table 5.4).

In Cambodia, sales of very small networks (fewer than 750 connections) during the rainy season are half what they are in the dry season. Larger networks experience about a 25 percent drop in sales in the rainy season. During the rainy season, 75 percent of households that had a network connection also used rainwater (figure 5.2). It is interesting that small numbers of households that are in the network area (but not connected) and households in the locality (but outside the network) used network water purchased from people with a connection.

Table 5.4 Uses of Water from Different Sources by People with Access to a Water Network in Benin, by Use, 2012

Percentage of respondents

	Source						
Use	Private connection	Standpipe	Well	River	Rainwater	Hand pump	Dam/pond
Drinking	100	100	71	23	79	100	..
Cooking	100	100	80	53	81	100	..
Laundry	100	100	100	100	100	100	100
Bathing	100	100	100	100	100	100	..
Garden	0	14	100	70	99	0	100
Watering animals	78	99	100	10	97	100	..
Economic use	100	98	100	100	100	100	..
Memo item							
People using source as percentage of people interviewed[a]	6	91	44	10	95	2	2

Source: Hydroconseil 2013.
Note: .. = less than 0.5 percent.
a. Some households use more than one source of water, so the percentages in this row add up to more than 100 percent.

Figure 5.2 Water Consumption in Cambodia in Dry and Rainy Seasons, 2012

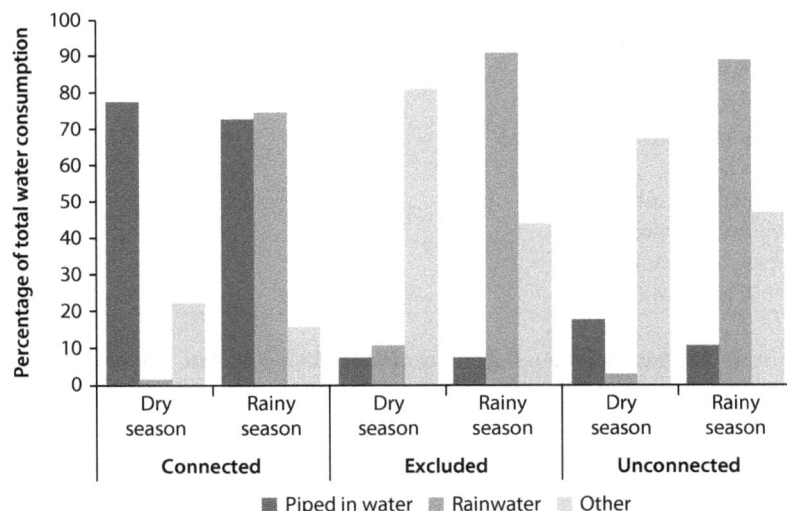

Note: Excluded = households outside the current reach of networks. Unconnected = households within the network service area but not connected to the system.

Service Features of Importance to the Poor

Poor rural residents are discerning consumers who make trade-offs among different sources of water and different expenditure items (such as mobile phones and water). Feedback from them suggests that firms need to provide a higher level of service if they are to convince these households to use (or use more) piped water.

Convenience and Reliability

In choosing one source of water over others, focus group participants in Bangladesh cited convenience, reliability, and quality (table 5.5). In Benin, convenience and quality were identified by 98 percent of respondents as being important factors in choosing a water source; significantly fewer people (78 percent) rated price as important (figure 5.3). An important dimension of convenience was proximity of the water source.

Table 5.5 Factors Influencing Choice of Water Source by Poor Rural Households in Bangladesh, 2012

Percentage of respondents

Factor	%
Only piped water supply (PWS)	
PWS more convenient	72
HTWs unsafe	64
No water in dry seasons	33
HTWs out of operation	26
HTWs located far away	24
HTW water tastes bad	18
Only hand pump tubewell (HTW)	
High tariff	53
Low service quality in PWS	29
Connection disconnected	24
PWS irregular supply	23
HTW convenient	10

Figure 5.3 Factors Influencing Choice of Water Source by Poor Rural Households in Benin, 2012

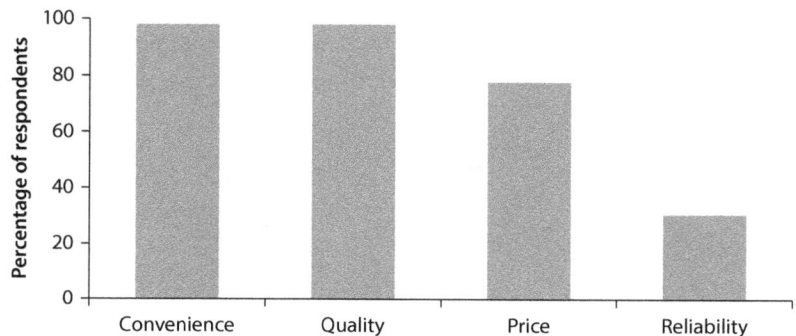

Figure 5.4 Consumer Satisfaction with Piped Water System in Bangladesh and Benin, 2012

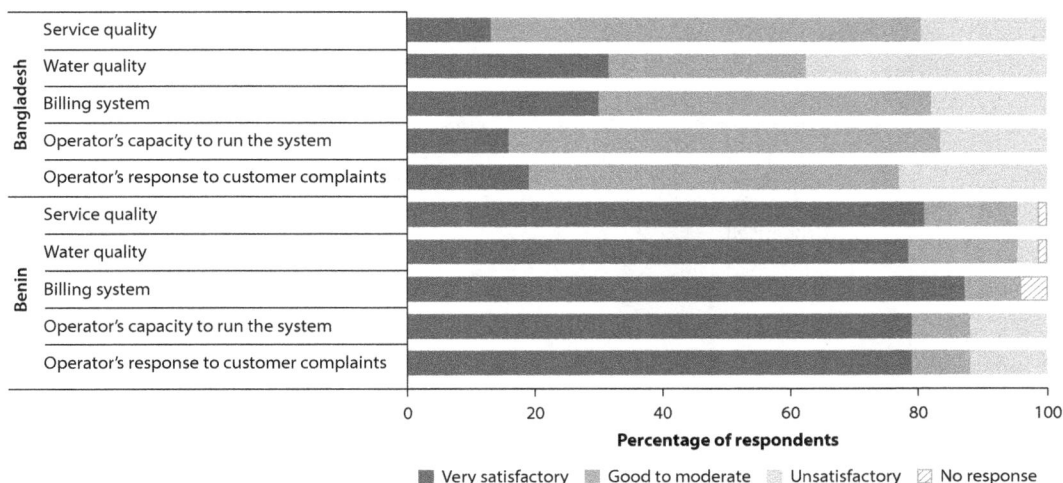

In Benin, more than three-quarters of respondents ranked their system and its operator as very satisfactory across all dimensions of performance, with just 12 percent rating response to complaints and capacity to run the system (in terms of ability to repair breakdowns) as being unsatisfactory (figure 5.4). Far fewer respondents in Bangladesh rated their systems as very satisfactory: across each dimension of performance, about a fifth rated it as unsatisfactory, and half of respondents with a connection said they would not recommend it to others, citing low pressure, irregular supply, and price relative to service.

Quality

People in all three countries rated the quality of piped water as high, although they focused on different aspects of quality. In Bangladesh, households associated quality with the elimination of the contaminants that made traditional water sources hard or unsafe to drink. In Benin and Cambodia, households were concerned primarily with cleanliness, texture, and color. In Cambodia, taste was an important factor; some households did not like the taste of chlorinated water.

Note

1. Per capita consumption of 40 liters a day is the World Health Organization standard. It is applied as a reference in the study countries. Households often purchase only part of their water consumption from networks, getting water from other sources, often at no monetary cost. In Benin, for example, the average household purchases about 4 liters per person per day; in Cambodia, network consumption averages about 35 liters per person per day. In Bangladesh,

where connected households pay a flat monthly tariff, consumption of network water averages about 85–114 liters a day, depending on the type of network.

Reference

Hydroconseil. 2013. *Benin: Deep Dive Analysis Report. Global Study for the Expansion of Domestic Private Sector Participation in the Water and Sanitation Market.* Cotonou: Hydroconseil.

CHAPTER 6

How Is Piped Water Supplied?

Enterprises building and operating networks share many characteristics, but they use very different business models, reflecting the incentives and opportunities created by the nature of government involvement. There is also considerable difference across the three countries in financial and service performance, which seems to be affected by the ways in which governments have shaped business opportunities.

Firm Characteristics

Firms operating networks are small. Some have licenses to operate networks, but only in Benin are they formally registered.

Firm Size

Rural piped water in Bangladesh, Benin, and Cambodia is provided largely by small businesses (operators with fewer than 20 employees) and microenterprises (operators with fewer than five employees) (figure 6.1). The sector also attracts some medium-size firms and firms that belong to larger enterprises.

In Cambodia, 70 percent of the enterprises interviewed were engaged solely in water supply as their primary business. In Benin, the proportion was just 44 percent; the majority of enterprises operated networks as a complement to their core activity (typically consulting services or contracting associated with other aspects of the water supply chain.) In Bangladesh, some schemes were operated as part of broader interventions by nongovernmental organizations (NGOs).

On average, enterprises in Bangladesh invested about $109,000 and enterprises in Cambodia about $90,500 in their networks (table 6.1). The investments made by enterprises are much smaller in Benin, where the public sector builds all infrastructure. As part of their application to manage systems, however, firms need to provide an inventory of their human and capital resources, which can include means of transport, tools, and office and communications equipment. Although some firms own nothing more than basic

Figure 6.1 Size of Operators of Piped Water Networks in Bangladesh, Benin, and Cambodia, 2012

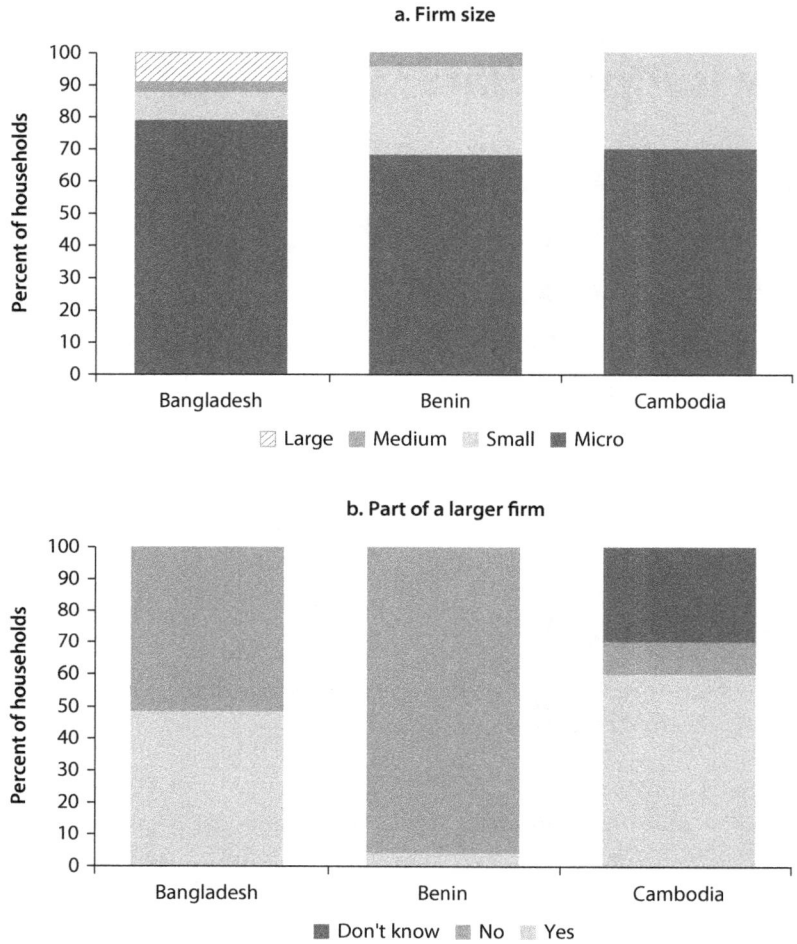

a. Firm size

Legend: Large, Medium, Small, Micro

b. Part of a larger firm

Legend: Don't know, No, Yes

Note: Size is based on number of workers, including working proprietors. Micro: fewer than 5 workers. Small: 5–19. Medium: 20–99. Large: more than 100.

Table 6.1 Total Investment by Water Network Operators in Bangladesh, Benin, and Cambodia, 2012

Dollars

Investment	Bangladesh	Benin	Cambodia
Minimum	6,900	200	1,900
Maximum	229,000	15,300	340,000
Average	109,000	2,800	90,500

furniture and equipment, more than half of the enterprises surveyed were operating more than one network, suggesting that their managerial capacity is more significant.[1]

Formality: Business Registration and License

The degree of formality of network operators varies considerably across the three case study countries.

In Benin, firms have to be formally registered in order to tender to operate networks. The enterprises surveyed were predominantly sole proprietorships or shareholding companies (figure 6.2). In Bangladesh and Cambodia, about 60 percent of firms were not formally registered as businesses, even though more than half of the enterprises in Cambodia had been operating for seven years or longer. Most of the registered firms in Bangladesh were NGOs.

As the provision of water supply is often regulated, the requirements of business registration and licensing are sometimes separated. In Bangladesh, there are no requirements for licensing. The authority to operate a water supply business is implicit in the cooperation framework under which the water infrastructure was financed and developed. In Benin, no separate licensing is necessary, because operators are selected through a formal tender process and in effect obtain their authority through the lease contract.

Cambodia has 139 licensed private operators of piped water schemes. The total number of private schemes (licensed and unlicensed) is estimated at about 300.[2] The number of licensed firms has increased over the years. Most firms interviewed for this study had applied for a license to operate a water network

Figure 6.2 Legal Status of Water Network Operators in Bangladesh, Benin, and Cambodia, 2012

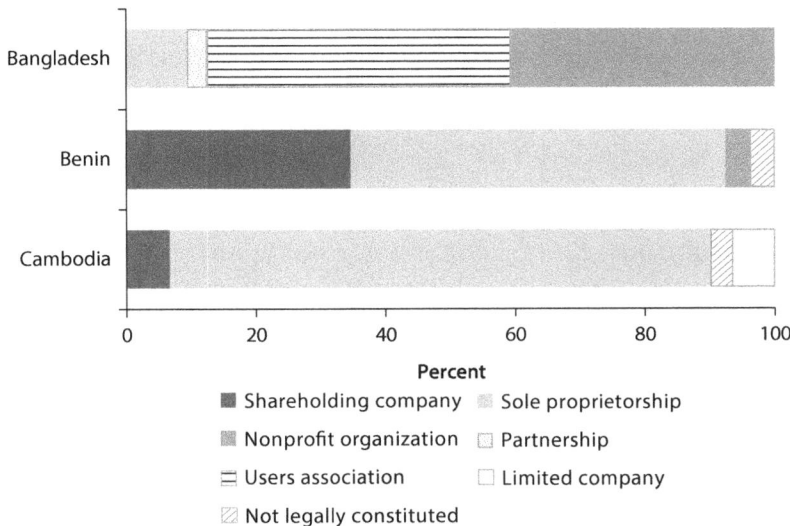

or were in the process of doing so. Firms that are licensed tend to share the following characteristics:

- They are larger than unlicensed firms, with 994 connections on average (compared with 254 for unlicensed firms), and they deliver significantly more water per connection.
- They make larger investments. On average, investment was three times larger than that of unlicensed firms operating schemes of comparable size.
- They have greater access to experts. A much larger proportion of licensed operations (69 percent) than unlicensed firms (7 percent) used external expertise to help with network design. Licensed firms were also more likely to have used the services of a construction company to build the network.
- They charge lower tariffs ($0.57 per cubic meter) than unlicensed firms ($0.61 per cubic meter) but higher connection fees (an average of $49, compared with $20 for unlicensed firms).
- They are much more likely to have a treatment plant. All licensed schemes had or were building a treatment plant. Only 53 percent of unlicensed schemes treat their water.

Business Models

The business models adopted by firms appear to be strongly shaped by the incentives and constraints created by government involvement. Typical models vary considerably across the three countries.

Revenue Generation

In Bangladesh and Cambodia, systems focus on smaller coverage areas, serving small numbers of customers with short networks (table 6.2). They sell large volumes of water through private connections. Networks in Benin cover twice as many people and the pipeline is twice as long, but they distribute water through standpipes.

Networks in Bangladesh sell more water than networks in Benin or Cambodia. Firms charge a flat rate per month, however. As a result, their revenues per cubic meter are just one-sixth those of their counterparts in Cambodia. Revenues per cubic meter sold in Cambodia are about half those of networks in Benin, but because sales volumes are much higher, revenues are twice those of their Benin counterparts. And while Cambodian networks deliver much less than networks in Bangladesh, their revenues are four times the size, given the difference in charging arrangements.

Tariffs and Pricing

Tariff levels vary widely within each country, but the variations do not seem closely related to differences in network costs. In Bangladesh, tariffs are set in consultation with the communities served, but it is not clear who is involved in this negotiation. There is considerable lack of clarity concerning ownership of assets. Partly because many networks are operated by NGOs and community-based organizations, tariffs are not set on a fully commercial

Table 6.2 Service Coverage and Revenues of Water Network Operators in Bangladesh, Benin, and Cambodia, 2012

Item	Bangladesh	Benin	Cambodia
Service coverage			
Number of villages served	3	4	6
Number of people served	1,504	8,023	2,403
Number of private connections	196	24	490
Number of shared connections	18	14	0
Average number of people per connection	7	216	5
Average network length (kilometers)	7	14	9
Revenues			
Average volume of water sold (cubic meters per year)	62,376	11,506	40,026
Average annual revenue (dollars)	5,200	10,700	22,200
Revenue per cubic meter of water sold (dollars)	0.08	0.9	0.5

basis. Eighty percent of firms interviewed in Bangladesh realized that tariffs do not cover the cost of the investment.

In Benin, tariffs are set using a model that also determines a renewal and extension fee, which is intended to cover depreciation of assets. Tariff rates are not used as a parameter for selecting firms tendering to manage a network but instead are preset by the *commune* (local government authority) based on the model. The average tariff in Benin is $1.03 per cubic meter—70 percent higher than in Cambodia and nine times as high as in Bangladesh.

In Cambodia, operators appear to negotiate tariffs with customers and with local government officials, although the central ministry responsible for urban water provides some guidelines. Ultimately, given the wide availability of alternative sources of water, tariff levels are determined by the market.

Asset Optimization

Most schemes in each country operate well below design capacity. In Bangladesh, average output was 18 percent of design capacity; in Benin the figure was 30 percent. Perhaps not surprisingly, in Cambodia, where private operators build their own networks, production was higher (46 percent utilization), albeit still well below capacity. On average, service hours were much longer in Cambodia than in the other two countries, and a smaller portion of schemes had unaccounted-for water rates that exceeded 20 percent (table 6.3).

The technical design and quality of construction of the main infrastructure components vary across settings. In Benin, where the central government built the networks, experts engaged for this study classified the pumps, the condition and protection of pipelines, and the protection of water sources in 80 percent of networks as optimal; in 20 percent of networks, the choice of pumping and storage facilities was considered suboptimal. In Bangladesh, where network construction typically had significant input from government or donors, a broadly similar pattern was seen. In Cambodia, more than a third of networks

Table 6.3 Performance of Water Networks in Bangladesh, Benin, and Cambodia, 2012

Indicator	Bangladesh	Benin	Cambodia
Average hours of service per week	33	86	132
Average water produced per resident served (cubic meters per year)	44	2	15
Percentage of networks with unaccounted-for water exceeding 20 percent	13[a]	9	4

a. Consumption is not metered; figure is estimate by study team.

were assessed as having suboptimal pump functionality and choice of pumping and storage facilities, and two-thirds were not providing adequate protection of the water source.

Financial and Cost Profile

To function sustainably, water networks need to generate enough revenues to cover operations, financing (where it exists), and depreciation. But only in Cambodia do operators face a clear imperative to cover the full value of network assets.

- In Bangladesh, the status of asset ownership is unclear. In most cases, there is no clear guidance on the level of provisioning for replacement.
- In Benin, the contract obliges the lessee to pay a fixed renewal and extension fee, which is based on a model that amortizes the construction cost of the water system. Fees covering replacement are paid to the local government (*commune*). The lessee also pays a *commune* fee to defray the cost of supervision. Where a water users association previously managed the network, the lessee may also pay a fee to the association. All of these fees are paid as fixed amounts per unit of water produced.
- In Cambodia, enterprises incur all capital costs themselves.

Financial Performance

In all three countries, about 80 percent of firms cover their operating costs (figure 6.3a). Only some of these firms also cover depreciation (figure 6.3b). In Cambodia, 90 percent of firms cover operating costs, and 80 percent cover their full costs. On average, operations cost more in Cambodia than in Bangladesh and Benin, but profit margins are also higher: average earnings before interest, tax, and depreciation were 48 percent of total revenue, compared with 43 percent in Benin and 31 percent in Bangladesh. Net profit margins were 23 percent in Cambodia and 17 percent in Benin; the average firm in Bangladesh incurred net losses of 45 percent.

Depreciation

In two of the three case study countries (Bangladesh and Benin), the majority of networks fail to cover depreciation. In Bangladesh, the problem is that

**Figure 6.3 Share of Water Network Operators Covering Operating Costs and Full
Costs in Bangladesh, Benin, and Cambodia, 2012**

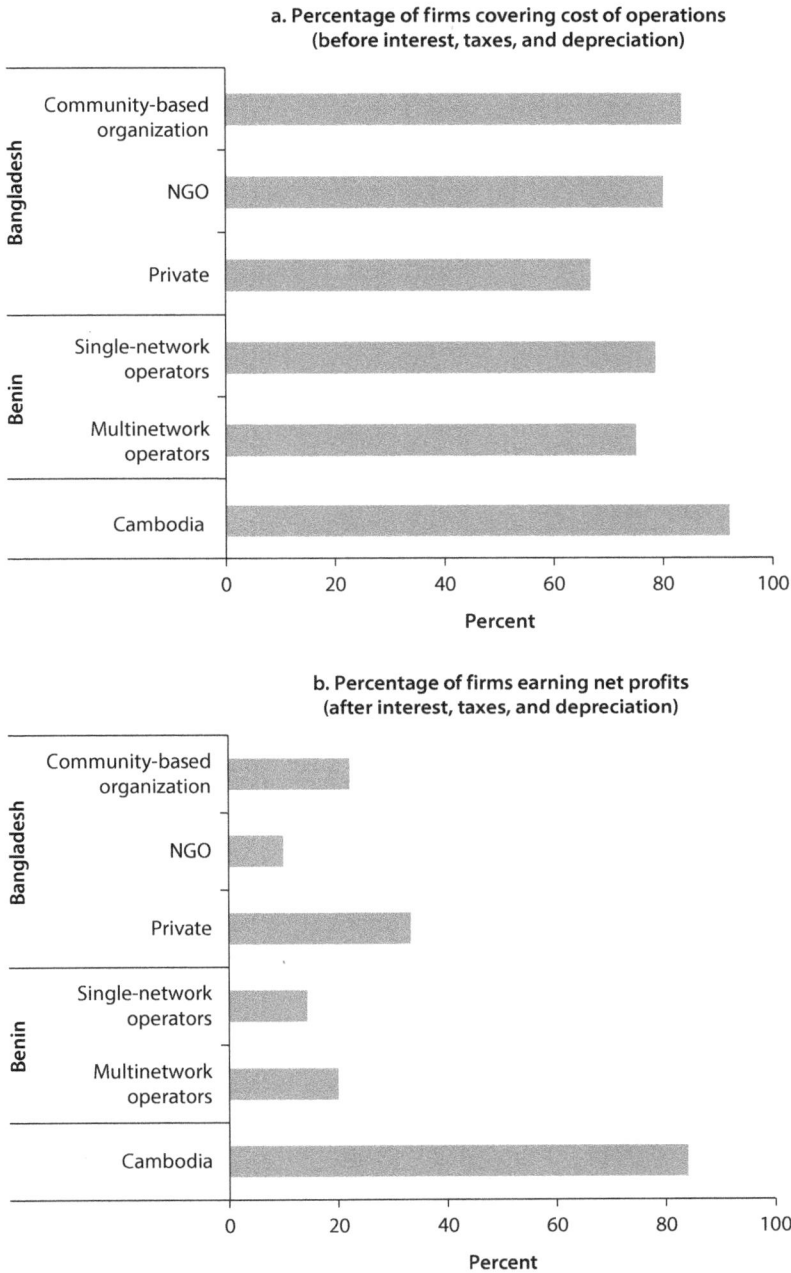

a. Percentage of firms covering cost of operations
(before interest, taxes, and depreciation)

b. Percentage of firms earning net profits
(after interest, taxes, and depreciation)

Note: NGO = nongovernmental organization.

Table 6.4 Sales, Cost, and Income Indicators of Piped Water Operators in Bangladesh, Benin, and Cambodia, 2012

Indicator	Bangladesh	Benin	Cambodia
Volume			
Water sold/resident (cubic meters)	41.5	1.4	16.6
Construction cost (dollars)			
Minimum	7,000	98,000	8,000
Maximum	220,000	960,000	340,000
Average (excluding outliers)	109,000	260,000	70,000
Construction cost/kilometer pipe (excluding outliers)	16,000	20,000	14,000
Estimated average annual depreciation	3,182	9,500	3,000
Operating cost (cents, except where otherwise indicated)			
Cost/kilometer of pipe (dollars)	512	631	1,643
Cost/resident served	240	113	362
Energy cost/cubic meter water produced	2	18	16
Variable cost/cubic meter water sold	9	57[a]	25
Tariff/cubic meter water sold	10[b]	103	60
Margin/cubic meter water sold	1	46	35

a. The variable cost in Benin, including fees, is $0.92 per cubic meter, leaving a margin for the operator of $0.10 per cubic meter.
b. Because customers are charged a flat rate, this figure is an implied charge per cubic meter.

effective tariffs are too low (table 6.4). As a result, after operating costs, the average firm is left with gross profits of just $0.01 per cubic meter of water sold, not enough to cover depreciation. Part of the problem is that firms charge customers the same amount regardless of the volume of water they consume, so that increasing consumption raises operating costs without increasing revenues.

In Benin, the problem is that depreciation costs, which are based on the costs of investment, are too high. These costs are high because networks are overdesigned and built to too high a standard, with expensive materials imported from Europe. The renewal and extension fee (which is meant to provide a return to cover depreciation to the local authority that owns the network) is calculated as a function of the cost of the underlying asset: the more a scheme costs, the greater the depreciation that has to be covered. The average cost of construction in Benin appears to be two or three times the cost in Bangladesh and Cambodia; on a per kilometer of pipeline basis, it is 25 percent more expensive than Bangladesh and 40 percent more than Cambodia. Networks in Benin are also too large for the markets they serve: they are designed to deliver 40 liters per capita a day, even though actual per capita sales average just 4 liters a day. Tariffs are calculated on the assumption that costs are spread over a much larger volume (and hence value) of water sold than is actually achieved. The problem is exacerbated by the fact that the fees are levied on the basis of production, not sales.

Energy Costs

Energy is one of the largest components of network operating costs (figure 6.4). It is particularly costly for operations that have to use diesel to power generators and pumps.

In Bangladesh, where all networks have fairly reliable access to the electricity grid, energy accounts for 39 percent of average operating costs. In contrast, in Cambodia, where the majority of networks use diesel to power pumps and generators, energy accounts for 65 percent of average costs; for networks wholly reliant on diesel, energy represents 74 percent of average costs. A similar pattern is evident in Benin, where the share of energy in operating costs is 20 percentage points higher for networks using diesel fuel than it is for networks using electricity. Pumping using diesel costs about three times as much as pumping with electricity.

Unreliability of power supply affects the ability of firms to deliver a consistent level of service to their customers. The problem affects all three countries. Ninety percent of firms in Bangladesh reported a power outage the previous year, with the average firm reporting six outages a year. In Benin, networks with

Figure 6.4 Breakdown of Cost of Water Network Operations in Bangladesh, Benin, and Cambodia, 2012

Percent

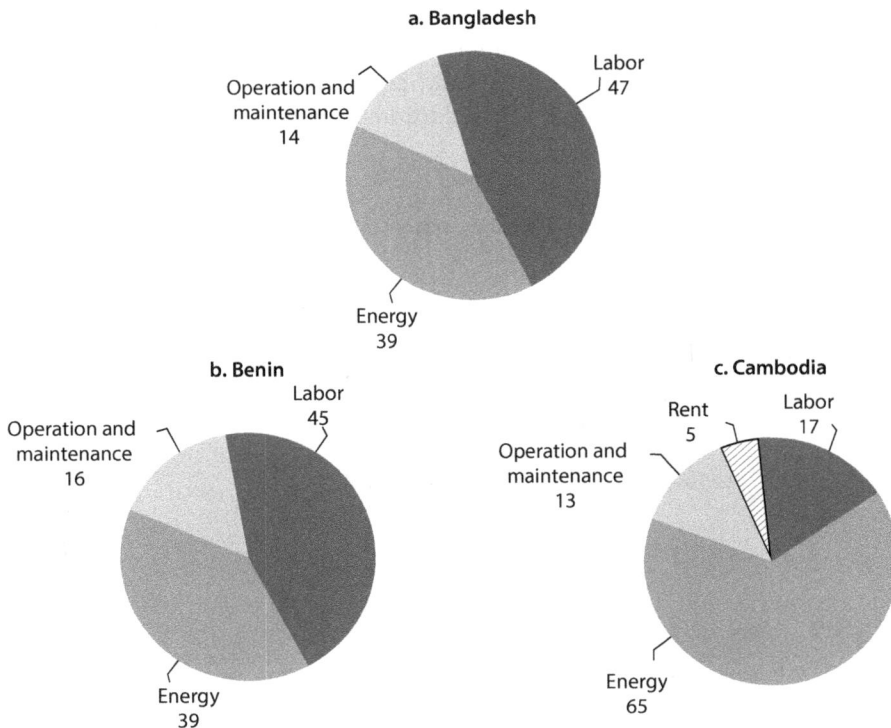

a. Bangladesh

Operation and maintenance 14
Labor 47
Energy 39

b. Benin

Labor 45
Operation and maintenance 16
Energy 39

c. Cambodia

Rent 5
Labor 17
Operation and maintenance 13
Energy 65

grid connections experienced outages of about 20 hours a month. In Cambodia, networks with connections to the grid experienced 30 outages the previous year.

Labor Costs

Labor costs largely account for the significant differences in variable unit costs across the three countries ($0.68 per cubic meter in Benin, $0.25 in Cambodia, and $0.09 in Bangladesh). Higher labor costs are associated with the operation of standpipes, through which most water in Benin is sold. Labor costs in Cambodia may understate true costs, because they may exclude some (unpaid) family labor.

Drivers of Profitability

In each country, country-specific drivers of costs and revenues interact to determine overall profitability (figure 6.5). It is in Cambodia, where the government has little direct involvement in network siting, design, construction, and operation, that firms and networks are consistently profitable and covering all costs.

In all three countries, there is a strong correlation between profit and various measures of network size, such as the length of pipe in the network and the number of residents served. In Bangladesh and Cambodia, where operators have some or complete control over the size and cost of the network, there is a strong correlation between size, investment, cost, and profitability (figure 6.6). Enterprises that operate more than one network do not achieve greater network profitability than enterprises with single networks, however. In Benin, the ability to spread overhead costs over more than one operation did not seem to compensate for the additional challenges of managing a larger

Figure 6.5 Cost and Revenue Patterns of Water Network Operators in Bangladesh, Benin, and Cambodia, 2012

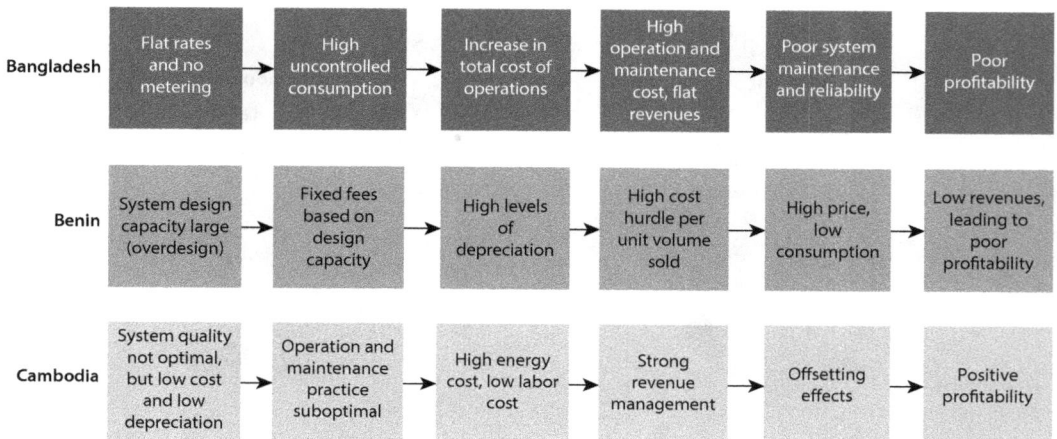

Figure 6.6 Correlation between Piped Water Operators' Net Profits and Investment in Bangladesh and Cambodia, 2012

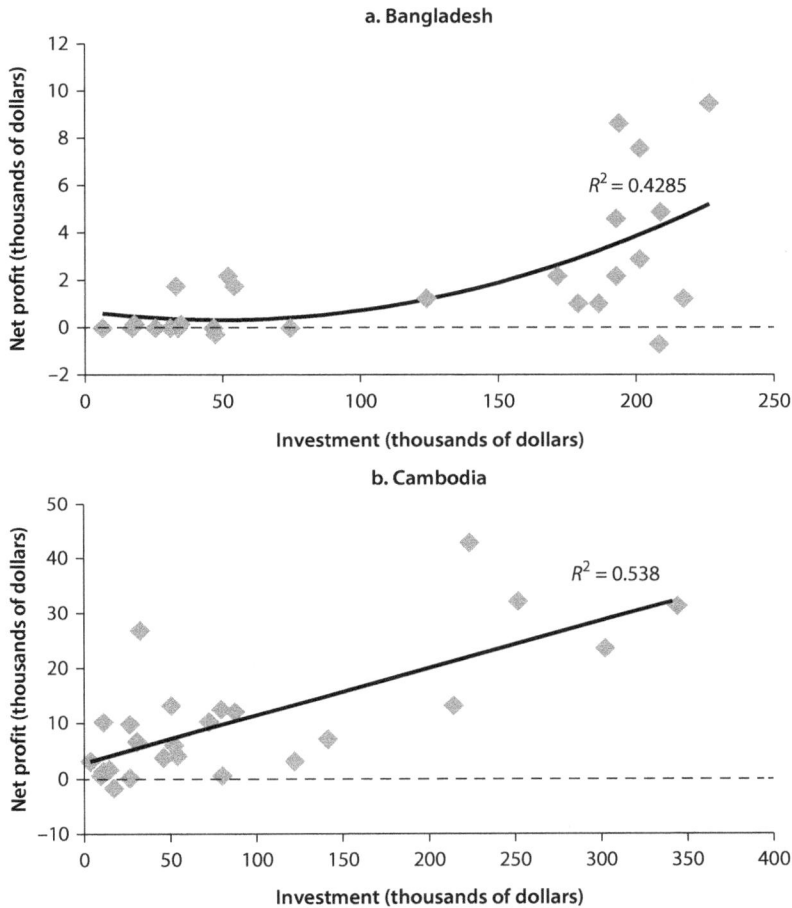

a. Bangladesh

$R^2 = 0.4285$

b. Cambodia

$R^2 = 0.538$

number of standpipe operators. Moreover, networks run by operators managing more than one network had above-average volumes of unaccounted-for water.

Private Connections

Across all three countries, the networks with the largest proportions of private connections are the most profitable. Consumption tends to increase with a private connection. Moreover, enterprises that use standpipes have to employ an operator to sell water and collect payment for consumption, not only adding costs but also creating principal-agent problems (a number of network managers in Benin perceive that theft of water facilitated by their standpipe operators is a significant problem). In Bangladesh, the number of connections is about 45–70 percent the number necessary for financial viability.

Tapping the Markets · http://dx.doi.org/10.1596/978-1-4648-0134-1

Figure 6.7 Private Connection Fees in Bangladesh, Benin, and Cambodia, 2012

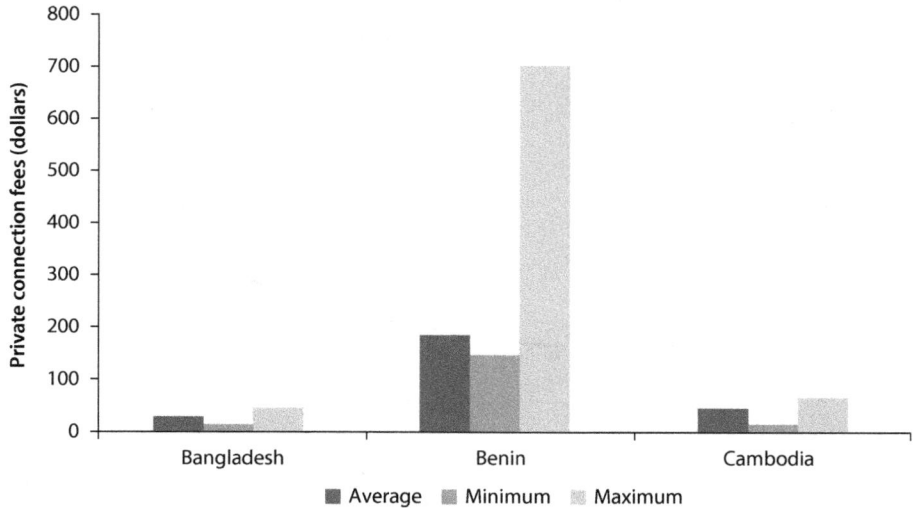

Note: Because connections are not metered in Bangladesh, costs may not be directly comparable to the other two countries.

The costs of private connections vary widely, both within and across countries. The average connection fee is nearly six times higher in Benin ($200) than in Cambodia ($34) (figure 6.7). One contributor to the difference is the fact that networks in Benin use much more expensive meters and connection materials (sourced primarily from Europe) than do networks in Cambodia, which use much cheaper products from China and Thailand.

Profit Motive and Alignment of Ownership, Investment Decisions, and Management

Having a profit motive and putting one's own money at risk are critical to actually earning profits. In Bangladesh, where piped water networks are sponsored and operated by a range of types of organizations, networks run by private enterprises that have invested in the network come closer to covering all costs than do government-sponsored networks or networks sponsored and operated by NGOs or community-based organizations. In Cambodia, where operators contribute all of investment capital, profitability is the norm. In contrast, in Benin, the agencies responsible for investment are not involved in operating the network, design is not aligned with market demand, and tariffs do not reflect market conditions. Networks cover larger numbers of consumers than their counterparts in Bangladesh and Cambodia, but they do so by providing lower levels of service (standpipes). Funds are not set aside for private connections, which are more profitable for the network, discouraging consumers from seeking them.

Interactions with the Construction Supply Chain

An important determinant of service quality and financial viability is the cost and quality of the water system. Both are affected by the supply chain for design and construction inputs. (Table A.5 in the appendix describes the supply chain in each country.)

In Bangladesh, firms did not report difficulty procuring materials or accessing technical services. Both supply chains seem fairly competitive.

In Benin, separate agents are responsible for building and operating networks. The national public investment program, under the management of the Department of Water, builds the networks, which are usually funded by donor programs. The process is lengthy, partly because it depends on national and international contractors and consultants. It is also expensive. Tender documents specify parts and components. Most pumps and generators are imported from Europe, whereas plastic pipes and connectors come mainly from the regional market, as do facilities for water treatment and most meters (box 6.1). In contrast, materials purchased during the operating phase, such as material for

---

Box 6.1 Construction of Water Networks in Benin

Construction of all water systems in Benin is handled through public procurement. The process is long and tedious, with the selection of a building company generally taking months or years. As most of the funding is external, funding agencies require the national government to implement stringent procedures to avoid leaks. These procedures further slow the selection process.

The capacity of both the Department of Water and the *communes* (local government authorities) is low. Drafting tenders, selecting and monitoring candidates, and paying them require specific skills that local agencies do not yet possess. As a result, the selection process is defective and the quality of the work undertaken by the building companies appears to be deteriorating, as illustrated by the rising number of breakdowns on new water supply systems.

Public procurement is not cost-effective: standards for equipment and parts set out in construction contracts are "oversized" in both quantity and quality given current demand for water. Most of the equipment and parts listed in the bill of quantities is imported from developed countries and, sometimes, specifically from the developed country that is financing construction of the system. These parts are generally expensive to purchase and even more expensive to import.

The market of informal water providers in peri-urban areas is more efficient. These providers invest their own money to provide water to identified customers. Technical standards are lower than on systems funded by the government. Equipment is directly or indirectly imported from Nigeria or China. The standard is lower, but the equipment is less expensive, enabling these informal providers to charge affordable prices to customers in their local area.

box continues next page

Tapping the Markets • http://dx.doi.org/10.1596/978-1-4648-0134-1

Box 6.1 Construction of Water Networks in Benin *(continued)*

The process for building water supply systems in rural areas through public procurement is not set up to respond to demand. Actors working in the water supply market pursue their own interests and objectives, which are not aligned with the demand for water:

- The Department of Water and local governments: Because contracting agents do not pay for the water supplied, they tend to require high standards and build assets that are larger than necessary.
- The contracted engineering consulting firm: The firm hired responds to the needs of the contracting agent (the Department of Water and local governments), which overestimate demand for water. The result is oversized water systems that cost more than water supply networks that are the right size.
- Donors: Donors' main objective is to supply funds. One of their measures of success is the amount of money they lend to governments. Funding mechanisms distort the supply of water infrastructure by directly or indirectly determining the origin, quality, and quantity of inputs. These choices are not optimal given the current use of water supply systems.
- Work contractors: Contractors are recruited to build, not to operate, the water supply system. To increase their revenues, they tend to push for larger systems. In addition, control of works is weak.

Source: Adapted from Hydroconseil 2013.

---

extending private connections (meters, pipes, and fittings) and operating the system (chemicals), are available through retail outlets selling imported products from China and Nigeria.

In Cambodia, private enterprises both build and operate water networks. The market for materials and equipment is well developed, with competitively priced and reasonably reliable material imported from China and Thailand. A problem is that few local companies are able to provide professional consulting services for design, construction supervision, or monitoring systems. International consultants with these capabilities are available, but they cater mainly to larger firms that build and operate urban water systems. Local enterprises cannot afford their services. Sixty percent of the enterprises surveyed for the study did their own design work, and three-quarters built network facilities themselves or used local tradesmen and laborers.

Notes

1. Of the 57 private firms managing water networks in Benin, 13 manage more than one network. One manages 27 networks, another manages 12.
2. Some sources cited in the Cambodia country study cited estimates as high as 800 (GRET 2013).

References

GRET (Groupe de Recherché et d'Echanges Technologiques). 2013. *Final Report Cambodia: Global Study for the Expansion of Domestic Private Sector Participation in the Water and Sanitation Market.* Phnom Penh: GRET.

Hydroconseil. 2013. *Benin: Deep Dive Analysis Report. Global Study for the Expansion of Domestic Private Sector Participation in the Water and Sanitation Market.* Cotonou: Hydroconseil.

CHAPTER 7

Are Firms Interested in Increasing Investment and Serving the Poor?

Eighty-nine water network operators were interviewed for this study. Their attitudes concerning expanding their businesses and providing piped water services to the poor suggest that opportunities exist for the domestic private sector.

Intentions to Invest

The policy framework within which firms operate shapes the investment opportunities open to them, and the business models that have developed as a result of that framework influences how they respond to these opportunities. Three-quarters of interviewed enterprises in Cambodia had intentions to invest, and just over half were interested in investing in an additional network (figure 7.1). At the other end of the scale, only a third of enterprises in Benin intended to invest, and only 30 percent were interested in operating an additional network.

In Bangladesh, investment in new networks is strongly determined by the availability of sources of complementary funding, primarily from the government and donors. Of the 55 percent of interviewed enterprises that were planning investment, all were focusing on expanding coverage of existing networks or repairing or improving performance rather than building a new network.

In Benin, the central government is responsible for building networks. Given the short duration of lease contracts, operators have limited incentive to invest in additional capacity. Moreover, given that the operator pays a replacement and extension fee to the local government that tracks depreciation of the original network, there is lack of clarity concerning the responsibility/incentive for replacing or upgrading investment. Forty-four percent of firms interviewed in Benin reported that they were planning investment, but this investment appeared to be maintenance spending to allow assets to continue functioning.

Figure 7.1 Enterprises' Intentions to Invest in Existing or Additional Water Networks in Bangladesh, Benin, and Cambodia, 2012

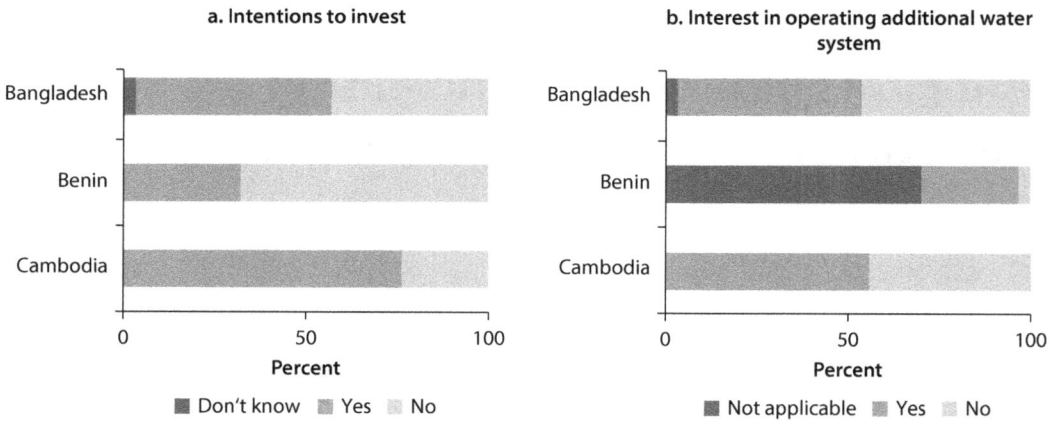

a. Intentions to invest

b. Interest in operating additional water system

■ Don't know ▨ Yes ▨ No

■ Not applicable ▨ Yes ▨ No

Figure 7.2   Areas for Future Investment Identified by Water Operators in Bangladesh, Benin, and Cambodia, 2012

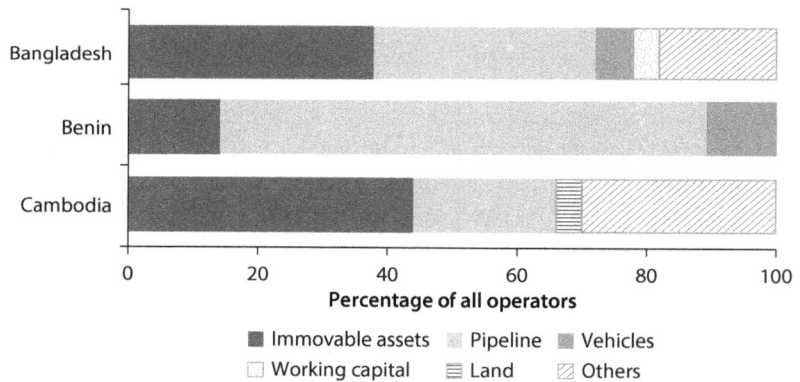

■ Immovable assets ▨ Pipeline ■ Vehicles
□ Working capital ☰ Land ▨ Others

In Cambodia, where network investment is autonomous and self-funded, enterprises face the full suite of potential investment options: expansion or enhancement of existing networks and the building of new networks. Just over three-quarters of enterprises interviewed were contemplating investments in existing networks, with a strong emphasis on expanding networks and water production; half of all enterprises indicated that they were interested in investing in new sites; and about a quarter were also contemplating other water-related investments, such as production of bottled water.

Figure 7.2 identifies the categories of investment that enterprises were thinking of making. Extensions to pipelines were the most frequently cited category. In Bangladesh and Cambodia, where enterprises own (fully or partly) the network, firms were also contemplating investing in network infrastructure.

Perceived Risks

Enterprises contemplating investment were asked about their perceptions of potential obstacles (figure 7.3). The findings, particularly in Benin and Cambodia, have important implications for policy that encourages firms to expand into challenging markets. They suggest that better market intelligence on potential markets is needed.

In Bangladesh, the ratings of obstacles for existing and additional systems were similar. In Benin, more firms rated the obstacles facing investment in new networks as very severe. Regarding new networks, firms were much

Figure 7.3 Obstacles to Investment in Existing and Additional Networks in Bangladesh, Benin, and Cambodia, 2012

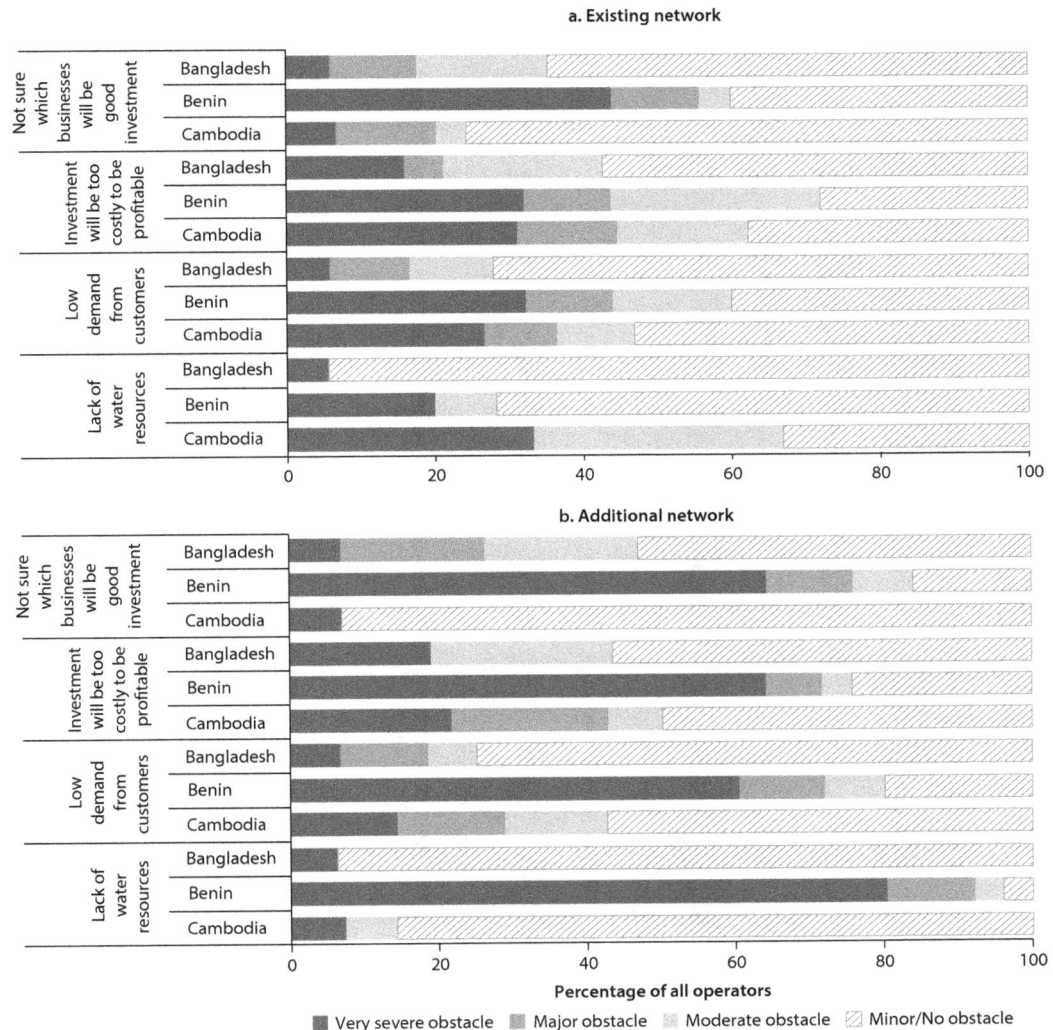

a. Existing network

b. Additional network

Percentage of all operators

■ Very severe obstacle ▦ Major obstacle ▒ Moderate obstacle ▨ Minor/No obstacle

more concerned with water availability, the strength of demand, the cost of investment and whether investing is a good business decision. They also expressed significant concerns about demand and investment costs in existing networks.

In Cambodia, many firms seemed to think that additional investments in existing systems would be too costly to be profitable, possibly because of low demand among people not currently served.[1] Incentives are likely to be needed to encourage firms to expand current networks to less profitable areas. Firms expressed slightly more optimism that there is greater demand in other systems and far more optimism that investment in other systems would generate profits.

Perceptions of the Poor as a Target Market

Enterprises in the three countries have quite different perceptions of the role of poor households in their markets (figure 7.4). In Bangladesh, just 30 percent of enterprises reported that the poor were their target customers, and 80 percent agreed with the statement that the poor did not have an equal chance to access their services. In Benin, 52 percent of enterprises considered the poor target customers, and 60 percent did not think they had equal chance of access. In Cambodia, 93 percent of enterprises identified the poor as target customers, and 80 percent reported that they did not have equal access to water.

At least 40 percent of operators contemplating investment in Benin and Cambodia (where all operators are commercial) were concerned about the profitability of catering to the poor. A majority of enterprises (58 percent in Bangladesh, 92 percent in Benin, 69 percent in Cambodia) cited the costs of

Figure 7.4 Enterprises' Perceptions of the Poor as Potential Customers in Bangladesh, Benin, and Cambodia, 2012

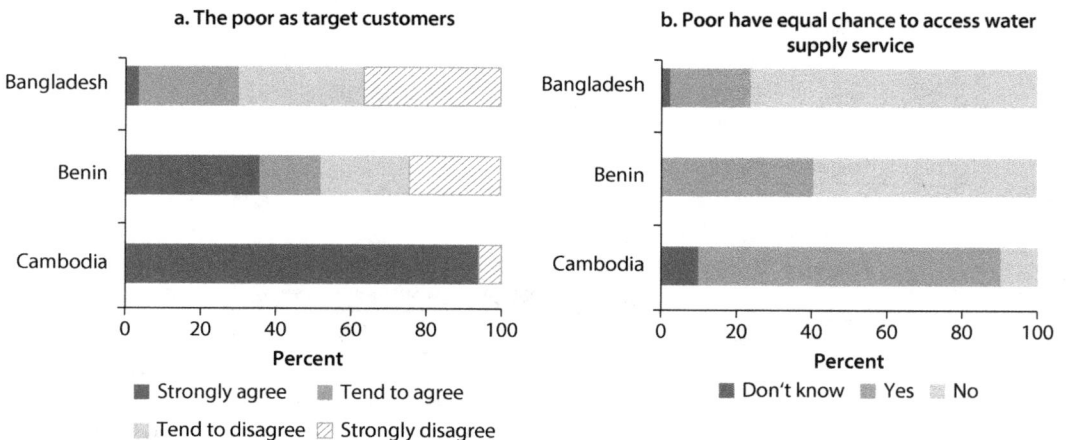

Figure 7.5 Enterprises' Views of Why Poor People Do Not Use Piped Water, 2012

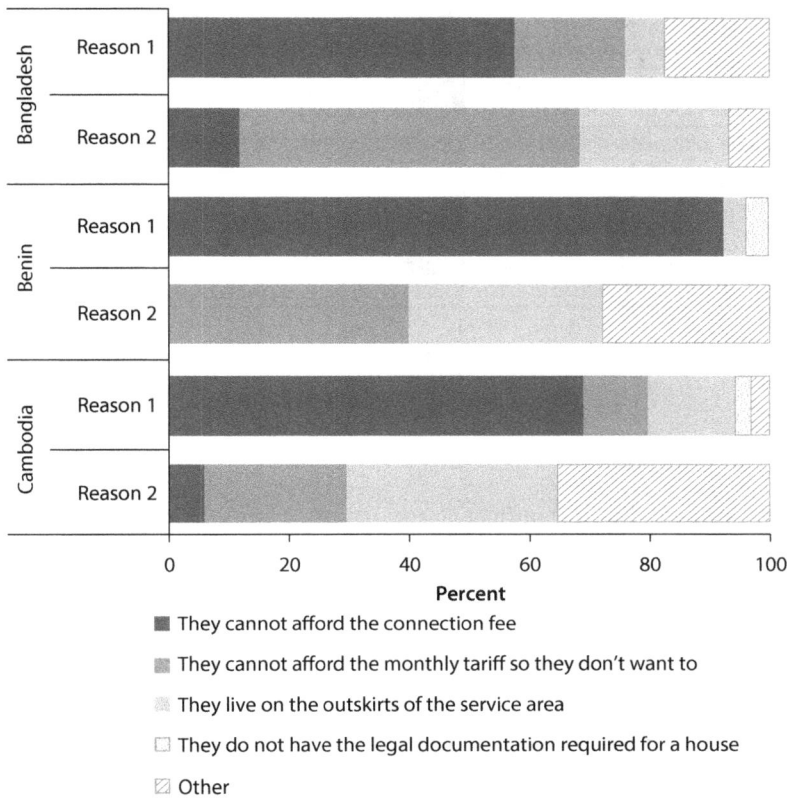

Legend:
- They cannot afford the connection fee
- They cannot afford the monthly tariff so they don't want to
- They live on the outskirts of the service area
- They do not have the legal documentation required for a house
- Other

Note: Enterprises in each country were asked to cite two reasons why they thought the poor did not have equal access to piped water.

connection as the main reason why the poor do not have equal access to piped water (figure 7.5).

A majority of enterprises in Benin and Cambodia think that poor households prefer cheaper water over good-quality water (figure 7.6). In contrast, a majority in Cambodia think that these households' demand for good-quality water is sufficient to justify catering to it. In Bangladesh, where quality is associated with the absence of dangerous contaminants, more than 60 percent of enterprises do not think poor households trade off price against quality, and nearly 90 percent think that demand for piped water by poor households is sufficient to justify catering to them.

Some enterprises in Cambodia view competition from other providers—water trucks, push carts, other piped water systems—as a possible constraint on demand for their services. But the critical challenge comes from self-provision and switching of sources. This finding highlights one of the fundamental challenges faced by piped water networks: they can deliver water at only one

Figure 7.6 Enterprises' Views on Water Preferences of Poor Households, 2012

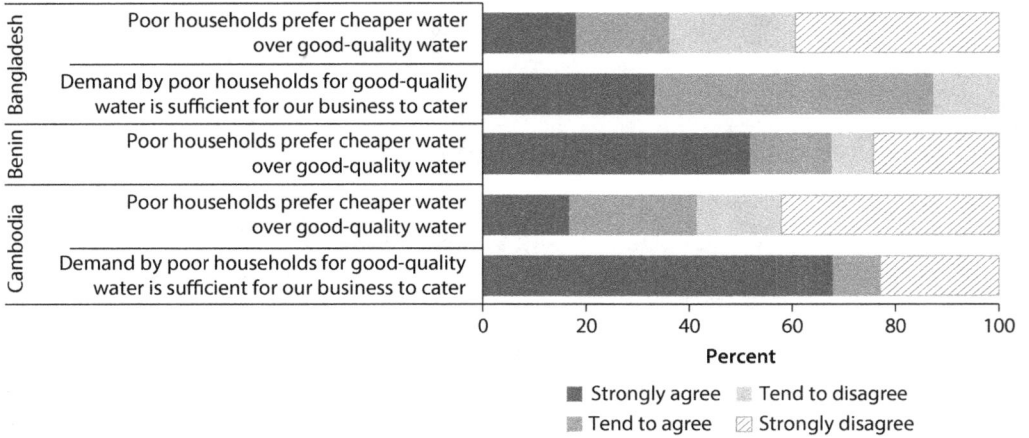

quality standard, but a significant proportion of household water use does not require that water be potable, and some uses do not even require that it be clean.

Note

1. The Cambodian firms' sense that the current market is near saturation is supported by a test conducted by the study team on potential returns to scale given current system configurations. It found that a doubling of production would increase costs by 88 percent. This finding suggests that although still potentially benefiting from scale economies, water systems in Cambodia are coming close to constant returns to scale. In contrast, doubling production would increase costs by an estimated 65 percent in Bangladesh and 79 percent in Benin.

CHAPTER 8

Is the Investment Climate Limiting Private Sector Involvement?

Government policy and practice, the quality of infrastructure, and access to finance shape the way firms perceive the trade-off between risk and return when considering expanding their business. All of these aspects of the investment climate affect the water sector.

Government Policy and Practice

Sectoral policies in all countries have deficiencies, and implementation and enforcement suffer from incomplete and under-resourced decentralization of responsibilities to local governments. Enterprise representatives were reluctant to criticize government policies and performance, however. Many seemed to accept certain undesirable aspects of public sector behavior, such as corruption (table 8.1).

In Benin and Bangladesh, market opportunities in water depend on investment made by the public (and donor) sector. Government policy is therefore a much more important issue for firms than it is in Cambodia.

Lack of Policy Specificity

In Bangladesh, the lack of specificity of the policy framework on engaging the private sector is a significant constraint. Business opportunities are limited to development projects that have set up clear rules of engagement. Consistency is lacking on tariff setting and financing rules and mechanisms. The ownership structures of the water business and its assets are not clear (most project documents state that the communities are the owner of the assets but do not define who the communities are or how they are constituted as a body). These problems stifle private sector initiative and increase opacity, preventing more competitive markets from developing. As a result, only quasi-private entities, such as nongovernmental organizations, have entered the market.

Table 8.1 Enterprises' Perceptions of Governance-Related Obstacles to Doing Business in Bangladesh, Benin, and Cambodia, 2012
Percentage of enterprises identifying issue as a problem

Country/severity of problem	Corruption	Competence of local government administration	Political instability
Bangladesh			
None/no view	33	15	33
Minor to moderate	21	67	55
Major to very severe	45	18	12
Benin			
None/no view	60	56	36
Minor to moderate	8	20	16
Major to very severe	32	24	48
Cambodia			
None/no view	77	100	97
Minor to moderate	13	0	3
Major to very severe	10	0	0

Note: Totals may not add to 100 because of rounding.

Lack of Capacity

In Benin, the main direct barrier seems to be the lack of capacity of the public sector in developing appropriate designed networks and of the commune to tender for their operation. But the broader barrier may lie in the fact that policy crowds out autonomous private investment similar to that which is made in Cambodia.

Within the current framework for building and operating networks, local governments, which are in theory responsible for developing and building water supply networks, do not yet have the capacity do so. As a result, arms of the national water administration system handle these functions. Operator selection has been reported to take as long as a year, and contracting construction can take years. The split responsibility for construction becomes more problematic when the network breaks down while still under the contractor guarantee or when technical information on the water supply network is required to set up the public-private partnership contract. The local government has little power to force the work contractor to carry out repairs and inherits the water supply network without specific documentation on the network (such as well yield, a map of the network, specific details of the electromechanical equipment), affecting the operations of the firm that takes over the management of the scheme. To overcome the weakness and delays associated with the fragmented responsibilities, the government might introduce other arrangements for engaging the private sector, such as design-build-lease contracts.

Lack of an Adequate Legal Framework

In Cambodia, the legal framework on urban and semi-urban water supply is still being developed. Regulation of private sector participation in water supply lacks clarity and exhibits a fair degree of informality and inconsistency in application. The licenses issued by the Ministry of Industry, Mining and Energy (MIME) authorize an enterprise to operate a network in a specific location and to use the associated water source for three years. Although this term is short, and could be seen as a deterrent to investment, enterprises seem confident that their licenses will not be revoked. Network operators appear to manage the short duration of their licenses by building infrastructure on land they own and by managing their relationship with local administrators.

Bureaucracy, Uncertainty, and the "Hassle Factor"

In addition to having to deal with the uncertainties and risk created by policy and the capacity of government institutions to implement it, firms face formal and informal costs of doing business in the sector. In Benin, for example, enterprises must be formally registered to be eligible to tender for water network leases, a process that costs $440–$590. In Cambodia, unregistered enterprises tended to believe that registration was not necessary or very expensive. The process for obtaining a license is not well documented or established by a legal document, but practice requires gaining official approvals at the commune, district, and provincial levels before MIME will issue a license. The informal transaction costs of acquiring a license ranged from $1,000 to $5,000, averaging just over $2,100; renewing a license averaged a little more than $500 every three years.

Piped water networks are immobile and relatively long-lived investments. The nature of the contractual arrangement that operators have with governments, enforcement of those arrangements, and mechanisms for resolving disputes are therefore critical in shaping incentives for investment. In all three countries, the World Bank's Doing Business surveys indicate that contract enforcement is a costly and time-consuming process (table 8.2).[1]

Difficulty Obtaining Land

The challenge of obtaining land on which to build networks makes it difficult for operators to expand. Along with the social capital that needs to be invested in local authorities, this constraint could be preventing firms from building networks in new areas despite their interest in exporting capital to other parts

Table 8.2 Time and Cost of Using Standard Legal Processes to Resolve Disputes in Bangladesh, Benin, and Cambodia, 2012

Country	Days required	Cost as percentage of value of claim
Bangladesh	1,442	63
Benin	795	65
Cambodia	401	103

Source: World Bank, 2013, Doing Business Indicators (database).

of the country. As the scope of the sector expands and competition for sites increases, more formal structures in licensing arrangements, provisions for land access, and dispute resolution mechanisms will be needed.

Operators in Bangladesh were skeptical about dispute-resolution processes. Formal arrangements for resolving disputes are not effective, and the weak quality of overall governance negatively affects the sector. Only 3 percent of respondents thought that the performance of their local government with respect to solving water supply problems was very satisfactory, and 38 percent considered it unsatisfactory. However, the fact that the government and donors are involved in most networks probably provides operators with a fair degree of comfort—in the short run at least—that they are cushioned against risks associated with unclear policies and legal protections.

In Benin, the rules and obligations of each party are stated in the standard lease contract; operators have to make a guarantee deposit, which can be forfeited if they do not meet the terms of the contract. In practice, however, breaches occur without consequence or are left to fester until a contract is terminated—as seen in the way breakdowns of the water system are dealt with. Contracts state that the enterprise is responsible for repairing all faults except of wells. This clause is not respected, and network operators do not recognize repairs as an obligation. When a breakdown occurs, the operator either repairs the fault and allows the costs to be deducted from the lease fee to be paid ex post or does not repair the system, which is then taken out of service. The operator rarely bears the financial burden of a breakdown.

Contracts in Benin do not grant exclusivity of service to operators within the area set in the contract or clearly provide protection against arbitrary termination of agreements by local councils. Operators cannot count on local councils meeting their commitments under contracts. However, most operators seem reasonably confident that their contracts provide assurance that their property, equipment, and other investments will not be arbitrarily taken by the state.

In this respect, the proactive supervision of water providers by local government matters. Some local governments seem better equipped and engaged in the monitoring and management of water providers than others. These local governments have generally appointed a full-time technical agent, who is mobile and has a computer.

In Cambodia, licensed operators have no legal guarantee of exclusivity of rights to provide and operate a network in a given commune. Nevertheless, half of respondents felt that licenses provided them with protection against competition from other suppliers.

Infrastructure

Firms in all three countries singled out unreliable power as the key infrastructure obstacle; the issue was most pronounced in Benin and Cambodia.

Energy is a critical input in Benin. Per cubic meter of water produced, fuel ($0.31) is more expensive than electricity ($0.12) However, connections to the grid are not universally available, and about 60 percent of rural water networks use energy from generators. Of networks that are connected to the grid, 40 percent have backup generators, which they use during power outages.

In Cambodia, 80 percent of the enterprises surveyed used fuel as the major energy source, purchased from one of four distributors. Sixty percent also purchase energy from the national grid operator, from private retailers selling energy from the grid, or from stand-alone private operators generating and distributing their own electricity. Fuel appears to be subject to nationwide pricing policies. In contrast, electricity tariffs vary considerably, ranging from $0.19/kilowatt hour (KWh) in Prey Vang Province to $1.00/KWh in Koh Kong Province. Where electricity costs more than about $0.43/KWh, operators tend to use fuel, generating their own electricity as necessary.

Access to Finance and Financial Services

Operators in Bangladesh and Benin have limited involvement in financing network construction. Lack of access to finance has therefore not typically been a significant obstacle; operators typically meet their expenditures from their own resources (figure 8.1). In contrast, in Cambodia, where borrowing from formal financial institutions is widespread, enterprises expressed concern about collateral requirements and (to a lesser degree) interest rates. Commercial banks require that loans be collateralized with land or buildings, typically with a value of at least 130 percent of the loan amount.

In Bangladesh and Benin, just over three-quarters of all firms interviewed had a bank account, but only 12 percent had a loan from a formal financial institution. In contrast, in Cambodia, only 17 percent of firms had a bank account, but slightly more had a loan (figure 8.2).

Figure 8.1　Access to Finance as an Obstacle to Investment in the Water Sector in Bangladesh, Benin, and Cambodia, 2012

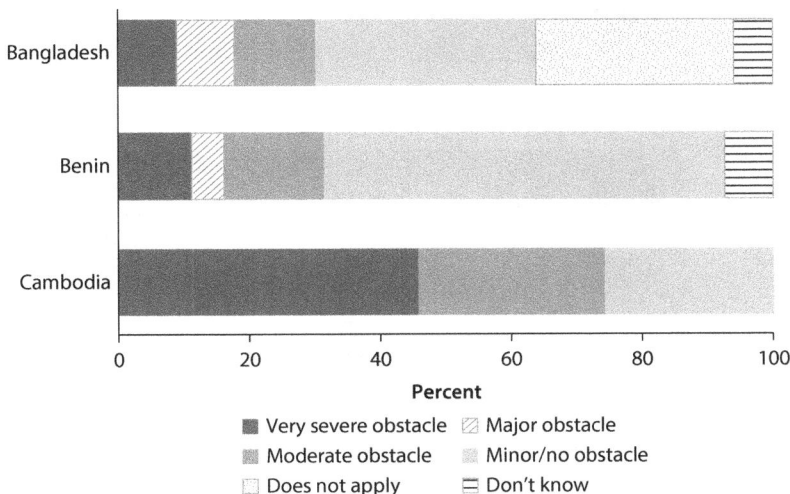

Legend:
- Very severe obstacle
- Major obstacle
- Moderate obstacle
- Minor/no obstacle
- Does not apply
- Don't know

Tapping the Markets · http://dx.doi.org/10.1596/978-1-4648-0134-1

Figure 8.2 Water Operators' Interactions with the Financial System in Bangladesh, Benin, and Cambodia, 2012

a. Prevalence of bank accounts

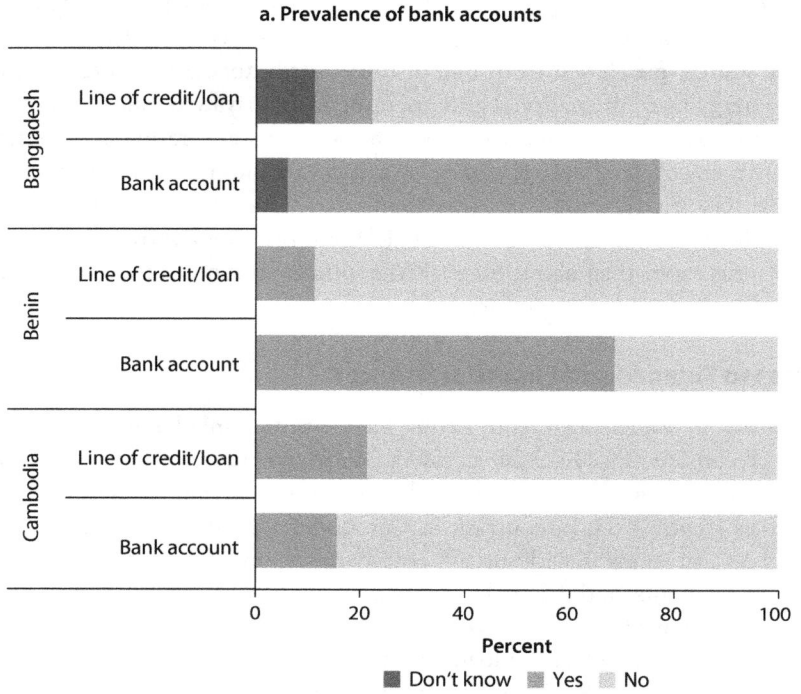

b. Sources of credit

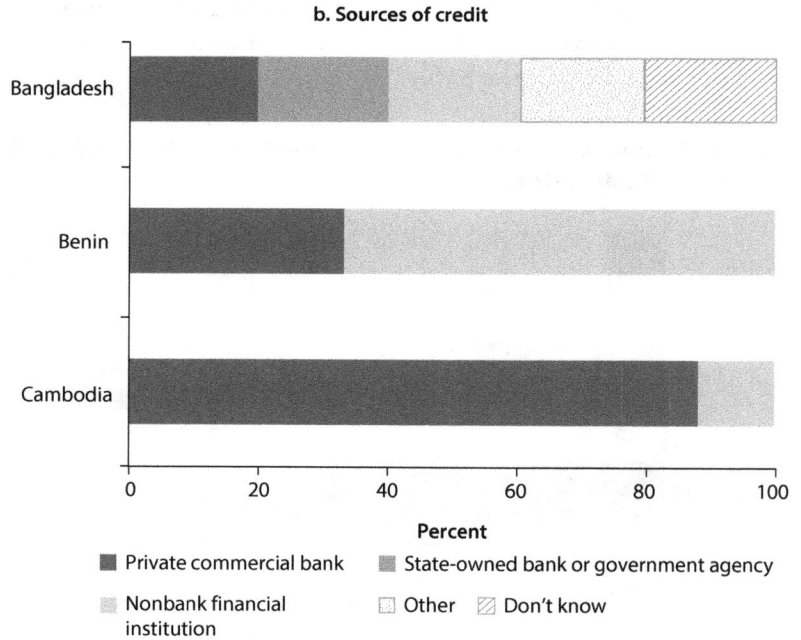

Note

1. The Doing Business project, managed by the International Finance Corporation and the World Bank, provides objective measures of business regulations and their enforcement for small and medium-size local firms in 185 economies.

CHAPTER 9

Conclusions and Recommendations

Only about 11 percent of the aggregate population of Bangladesh, Benin, and Cambodia gets its water from state utilities—the rest rely on a combination of self-supply, private provision, and community-run systems. As population densities increase, piped water networks are becoming a viable source of water supply in many more settlements, and the private sector is becoming increasingly involved in the construction and operation of nonurban networks.

A variety of constraints prevents the domestic private sector from increasing supply from these networks to the poor, however. Commercial, water policy, and investment climate conditions all play roles in limiting private provision. By improving these conditions, donors and government policy makers could make the delivery of piped water to poor people in these countries more attractive to private investors.

Conclusions

The study aimed to characterize the potential market for rural private piped water systems and to answer the following questions:

- Is lack of interest by the domestic private sector a rational response to weak market potential, or are lack of firm viability and the use of inappropriate business models preventing it from taking advantage of market opportunities?
- Are policy and investment climate factors increasing the (actual or perceived) cost and risk associated with doing business?

Market Potential Is Strong

In the three study countries, the potential market the domestic private sector could be serving is large. Projections suggest that by 2025, about 20 million people in Bangladesh, Benin, and Cambodia will get their water from rural piped water schemes—10 times the current number. This market will be worth at least $90 million a year, up from about $23 million in 2012.

Market growth is being driven by a combination of economic and policy factors. Population and income growth are important, but country-specific drivers are at play as well:

- In many locations in Bangladesh, current sources are unsustainable, because of contamination and the growing scarcity of water. A national policy aims to respond to these problems through public/private/community coinvestment in piped water networks.
- In Benin, the government recently adopted a policy to contract out management of networks built by the public sector.
- In Cambodia, the costs of alternative sources, the absence of public supply, and a liberal (if somewhat unregulated) government approach to licensing private networks are creating commercial opportunities.

Constraints Are Preventing Private Investors from Serving the Poor
Commercial factors are broadly similar across the three case study countries. Policy factors are more country specific. Although all three countries recognize the role of the private sector in increasing access and improving quality of service, each has policies that make it difficult for private firms to be profitable, thereby dampening their interest in investing.

Weak demand. Households, especially poor households, purchase too little water from networks for operators to achieve optimal capacity utilization or to warrant significant investments in additional capacity. Poor households need higher volumes of water, but their purchases are limited by cost and their assessment of the value of network water with respect to alternatives.

Most households have access to inexpensive alternative sources of water (if only for parts of the year), including wells, springs, and boreholes. They are savvy about making trade-offs between price and value in choosing their water source. In the longer run, the availability and opportunity cost of alternatives will likely shift incentives in favor of networks, especially if operators can assure consumers of the quality of the service they offer. In the short run, however, competition from other sources will limit demand for piped water.

As a result, firms struggle to cater to the poor. To provide an acceptable standard of service—in terms of water quality, convenience, and reliability and continuity of supply—and cover network costs, they need to make good use of their fixed assets by selling as much water as network design will allow. Private connections appear to be an effective way to encourage households to increase purchases of piped water, but they are expensive relative to the incomes of poor households. There are currently no broadly applicable approaches to addressing issues of affordability or financing of outlays for water connections and services for poor households that are consistent with financial sustainability.

Private operators also face challenges in signaling to customers that their water is consistently safe to drink. Where public administrations are capable and corruption is not rampant, countries typically address the problem through state accreditation of the processes and systems used by utilities, accompanied by occasional testing and audits. National capacity to deliver regulatory functions has been challenged by the extent of decentralization of responsibility for water sector management: in Benin and Cambodia, decentralization has not been realized because it has not been accompanied by appropriate efforts to develop capacity and ensure adequate resourcing of devolved functions. The sector and governments need to develop ways to educate consumers about water quality and enable networks to adopt standard approaches to ensuring it.

Lack of firm viability and inappropriateness of business models. At a certain network size, piped water systems offer considerable economies of scale in providing potable water. But reaping these economies requires operating above certain minimum levels of sales, and economic and financial sustainability requires charging prices that cover all costs. Getting this balance right is challenging. Firms in Cambodia typically recover the full costs of investment. In contrast, their counterparts in Bangladesh and Benin seem to struggle to maintain a reliable level of supply that meets customers' expectations.

Different business models have emerged in the three countries as a result of market and policy drivers. Each has achieved a different degree of success.

In Bangladesh, private sponsors coinvest in networks with the government and donors in localities where groundwater cannot be safely used. Customers are served through private connections. They pay a flat monthly fee, which results in low revenues despite high volumes. Combined with the fact that most networks have too few connections given the investment cost to households, the tariff structure means that few networks are financially viable.

In Benin, the business model is to cover a larger service area through manned standpipes. A top-down investment program designs and builds all networks, which are too large given the scale of the market. The tariff structure is determined by policy-driven financial models that grossly overestimate market sales, leading to very high fees and tariffs. Tariffs provide a large profit margin for every unit of water sold—and most operators therefore make a profit on their leases—but they keep consumption levels low. As a result, aggregate revenues do not cover investment costs.

In Cambodia, financing, design, construction, operation, and management are wholly private. Networks serve households through metered connections. Nearly all networks yield positive returns on investment, and revenues enable adequate provisioning for depreciation. Designed capacity is well calibrated to the market, and continuity of service is good. But lack of access to water sector expertise may be leading to suboptimal choices of design and equipment, and the potability of water may not be ensured.

Attitudes toward investment and serving the poor. Water firms in Bangladesh and Benin, where the public sector and donors largely determine which assets are built, are circumspect in their attitudes toward investment. Few firms in Bangladesh were planning investment, and the investment that was planned focused on expanding the coverage of or repairing existing networks. In Benin, nearly half of firms interviewed were planning investment, but spending seemed to be going toward maintenance to allow assets to continue functioning. In contrast, in Cambodia, three-quarters of enterprises interviewed were contemplating investments in existing networks, with a strong emphasis on network and water production expansion, and half of the enterprises were interested in investing in new sites.

Enterprises identified a range of market-related risks that affect their investment plans. In Bangladesh, the main concern was that costs make profitability uncertain, for both existing and new networks. This concern reflects current conditions in the market, where most firms are not profitable. In Benin, enterprises cited their lack of experience in developing (as opposed to operating) systems. They cited a wide range of risks, including concerns about water availability, lack of sufficient demand, and high cost of investment, and expressed uncertainty about which investments to make. In Cambodia, firms' greatest concern was access to finance. Cambodian firms display a strong orientation toward serving the poor. In contrast, few firms in Bangladesh or Benin considered the poor as their target market, and many believed that their policies did not provide the poor with equal access to their service. In all countries, firms believed that costs are beyond the reach of the poor and that no incentives exist to reach these harder markets.

Unsupportive investment climate. In addition to market-related risks, firms face a variety of policy and institutional obstacles. In Bangladesh, the pricing and ownership structures do not seem to allow networks to recover—or even earn a return on—their capital costs, and investment is contingent on government or donor cofinancing. As a consequence, private operators appear reluctant to expand networks or sponsor additional networks. In Benin, the main barrier to expansion is the lack of capacity of the public sector in designing appropriately scaled networks and tendering them for private operation and the nature of the leases under which firms operate networks. In Cambodia, the incomplete nature of the legal framework on urban and semi-urban water supply and lack of clarity and consistency about the rules governing private investment in water networks may be constraining the types of investment that private firms are prepared to make. The challenges associated with acquiring land also dampen operators' interest in investing.

The lack of good physical and financial infrastructure also stifles investment. Firms in all three countries singled out unreliable power supply as a key constraint to doing business. Energy is a large element of operating costs, accounting for 39 percent in Bangladesh and Benin and 65 percent in Cambodia. Where networks use diesel fuel (to generate electricity full time or as a backup or to run intake pumps), energy costs are significantly higher.

The limited reach of the financial sector and the costs of accessing finance also limit firms' ability to invest. In Cambodia, for example, all loans must be collateralized by real estate.

Recommendations

Water sector policies in all three countries recognize the contribution that private provision plays in meeting household needs. But current policies typically still work from a starting point of modifying models of state delivery rather than enabling poor households to interact with private systems. Much more could be done to stimulate the development of efficient markets in which private suppliers are motivated to meet the needs of the poor.

The study offers recommendations in three key areas: stimulating the demand for network services by the poor, improving business viability and business models by removing policy and other impediments to efficient behavior by private firms, and improving the investment climate and the incentives to invest in challenging markets (table 9.1). Because the prevailing models in the three countries span a broad spectrum of approaches, some recommendations are more relevant in some cases than others. They need to be adapted to specific country circumstances and the overall direction of policy toward the water sector.

Table 9.1 Policy Recommendations for Increasing the Provision of Piped Water to the Poor

Policy goal	Recommended action	Actor
Stimulate demand by the poor		
Improve affordability	1. Develop the right size:	Governments, development partners
	• Design and build assets that are appropriate for small-scale networks, so that cost-recovery prices can be kept as low as possible.	
	• Realistically assess demand and adopt standards and procurement rules to align network design with it.	
	• Modify tendering systems to identify inputs in terms of performance and quality standards rather than by specifying particular brands or suppliers.	
	2. Smooth and subsidize expenditures:	
	• Experiment with initiatives that enable poor rural households with volatile cash incomes to spread connection payments (and perhaps usage charges) over time.	
	• Where facilities for cash transfers to the poor already exist, consider providing targeted demand-side support for the extreme poor.	
	• Where networks are leased to private operators or involve coinvestment by government or donors, consider building into lease contracts or project designs a requirement to offer concessional terms for connections to poor households.	
	• Where network construction and operation are completely independent of government and donors, consider delivering support directly to households, rather than trying to impose community service obligations on operators.	
	• Develop financing schemes that enable operators to offer customers installment plans for paying for private connection costs.	
Establish appropriate standards	3. Help firms signal water service and quality to the market:	Governments, development partners
	• Identify service and quality service standards and means of achieving them that are both consistent with regulatory capacity and simple enough for consumers to understand.	
	• Help firms to implement procedures for ensuring water quality and to target information campaigns to their customers.	
Improve business viability and business models		
Improve profitability	4. Remove impediments to efficient pricing, without which private operators cannot be financially viable:	Governments, development partners
	• Introduce metering, so that firms are paid for increased usage (Bangladesh).	
	• Where tariffs and charges are regulated, recalibrate models to avoid setting tariffs so high that they restrict consumption excessively (Benin).	
	5. Optimize the operations of the network under contract, where contracted-out networks face competition from other publicly owned water sources.	
	• Assess the feasibility of regulating exclusivity and alternative delivery in network locations (by including public water points in operator contracts with appropriate pricing, for example).	
	• Develop regulated arrangements for sharing connections or resale of water to increase consumption and capacity utilization.	

table continues next page

Table 9.1 Policy Recommendations for Increasing the Provision of Piped Water to the Poor *(continued)*

Policy goal	Recommended action	Actor
Expand private connections	6. Establish incentives for incremental upgrades of existing networks to offer more private connections, which provide the convenience consumers strongly value: • Grant concession contracts or enhanced lease contracts in which the private operator implements publicly funded investment in network expansion/densification (Benin). • Improve the planning, marketing, and design of networks so that water points are located where households need them, and promote the use of private connections (Bangladesh and Benin).	Governments, development partners
Improve supply chains and technical support	7. Improve professional capabilities for the design, construction, and maintenance of small-scale piped water networks: • Foster the creation of professional associations to train and provide accreditation for consultants who design networks or provide other expertise to small-scale water operators. • Support business brokering initiatives that could work with financial institutions to assess the risks and feasibility of network investments by small enterprises. • Reduce the size of lots in the public procurement of water infrastructure development, in order to allow local players to compete and build capacity.	Governments, development partners, the business community
Improve the investment climate and sectoral policies		
Provide market intelligence	8. Improve information for potential investors about investment options, so that enterprises are aware of the availability of water resources and market potential in areas outside their current locations of operation: • Improve sector investment planning to identify—and publicize—markets with potential for private participation. • Provide technical support to local authorities to develop projects that can be taken to market.	Governments, development partners
Increase access to finance	9. Address the low level of financial inclusion and the limited availability of financing for small water projects: • Develop financing facilities to support cash flow–based financing for water projects, including the use of blended funds, guarantees, and cost-sharing arrangements, and provide appropriate project development and appraisal support to financial institutions. • Develop robust loan documentation that is consistent with national legal frameworks, and assist with legal reform and clarification to facilitate market-based financing of and investment in water projects.	Governments, development partners
Increase access to land and energy	10. Facilitate access to land for private water schemes, and address the high cost and limited and unreliable supply of energy: • Where concession law structures are in place, use them to bring small-scale water projects to market with provisions for land access and infrastructure development (Cambodia). • Consider offering incentives for generating power for water projects in locations that are poorly served by the grid.	Governments, development partners

table continues next page

Table 9.1 Policy Recommendations for Increasing the Provision of Piped Water to the Poor *(continued)*

Policy goal	Recommended action	Actor
Improve government policy and practice	11. Improve policy clarity and functionality to facilitate provision of piped water in more marginal locations: • Prepare operational guidance on the role of the private sector, and move from project- to policy-based approaches to increase transparency and competition and avoid distortions created by inconsistency and idiosyncratic subsidization (Bangladesh, Benin, Cambodia). • Improve arrangements for determining tariffs, and introduce incentives for expanding coverage and meeting service standards (Benin). • Where the prevailing model is public–private partnerships, improve incentives for sustainable service delivery by including incentives to expand coverage and meet service standards; improving arrangements for determining fees paid by network operators; tying them to likely revenues and costs; and clarifying responsibilities for repair, replacement, and expansion of the network (Bangladesh, Benin). • Where the prevailing model is autonomous private investment, develop a system of competitive tendering of rights in hard-to-reach or less profitable localities using a more traditional public–private partnership model, and ensure that interventions that stimulate private provision create a level playing field (Cambodia). 12. Strengthen dispute-resolution arrangements, the absence of which deters investment: • Provide training programs for public and private parties to contracts to improve their understanding of obligations, and introduce mechanisms to support regular business planning and performance review processes as a companion to dispute-resolution arrangements. • Empanel independent reviewers and auditors to help contracting parties resolve disputes.	Governments, development partners

Sanitation

Overview of the Sanitation Sector

Throughout the developing world, millions of people lack access to improved sanitation. In the four countries covered in Part 2 alone, the problem affects some 228 million people and costs 1.0–6.3 percent of gross domestic product (GDP)—a total of at least $10 billion a year.

Part 2 examines private sector provision of on-site sanitation services in Bangladesh, Indonesia, Peru, and Tanzania, four countries where the local private sector already plays a major role in helping rural (and many urban) households construct and maintain sanitation facilities. In Bangladesh, Indonesia, and Tanzania, at least 95 percent of the population with some kind of toilet relied on a private initiative to construct their facilities. Even in highly urbanized Peru, where public utilities have long provided sewerage systems, a quarter of people with some kind of sanitation use privately constructed latrines/toilets and septic tanks. Little systematic information is available about these markets; most information on the private sector in sanitation focuses on large private enterprises that provide wastewater management services.

Bangladesh, Indonesia, Peru, and Tanzania are countries where the Water and Sanitation Program (WSP) is actively supporting client governments in engaging the domestic private sector. The WSP—a multidonor partnership administered by the World Bank to support poor people in obtaining affordable, safe, and sustainable access to water and sanitation services—is well placed to offer practical follow-up of the study results in these countries.

In each country, the study examines the preferences and circumstances of poor households and the performance of enterprises that provide sanitation-related services directly to them. It examines commercial and investment climate factors that may affect enterprises' actual or perceived costs and risks, driving their decisions about increasing investment in their business. Specifically, the study seeks answers to the following questions:

- Is lack of interest by the domestic private sector a rational response to weak market potential, or are lack of enterprise viability and the use of inappropriate business models preventing the private sector from taking advantage of market opportunities?
- Are investment climate factors increasing the (actual or perceived) costs and risks associated with doing business?

Market Potential for On-Site Sanitation Services

The current market for improved on-site sanitation services in the four countries is large: supplying new systems and replacing old ones is conservatively estimated to be worth $300 million a year. But the potential market is much larger: one-time sales of improved sanitation facilities to the 228 million people without access are worth at least $2.6 billion. Poor people alone would account for sales of about $700 million. New customers would increase the replacement market to about $550 million a year.

Private sector activity associated with the market is not limited to the installation of latrines and toilets. The domestic private sector in these countries is engaged in a range of activities, including wholesale and retail sales of materials and components, the manufacture of prefabricated cement products used to build latrines and toilets, and the provision of advice on and the design of latrines and toilets. Some enterprises also offer financing facilities or are engaged in related services, such as repairs, pit emptying, and septage disposal, which could represent sizable business opportunities (the potential market for truck-based pit emptying in Indonesia is about $100 million a year, for example).

Constraints to Serving the Market

The main constraint to the scaling up of private sanitation to the poor and realization of the market's potential is the fact that enterprises are not offering households products and services they want to buy. Many poor (and not-so-poor) people are unwilling to pay for the kinds of improved sanitation solutions currently available. As currently structured, the supply chain delivering these solutions appears unable to offer better value.

Weak Demand for Existing Options

Sanitation is a low expenditure priority for poor households. Cost is an important factor, but it is not necessarily an insurmountable barrier. The improved on-site sanitation options currently available cost between 3 percent (Bangladesh) and 7 percent (Peru) of the annual income of poor households. Many poor households spend considerably more on consumer durables such as mobile phones. In Bangladesh, for example, 100 percent of poor families living on between $62 and

$122 a month in the areas covered by the study had at least one mobile phone, as did a third of extremely poor families living on less than $62 a month. Average annual outlays on phones amounted to nearly twice the cost of a standard improved latrine or toilet.

Households do not purchase improved sanitation because they do not find current options attractive enough. Poor households are faced with limited options and significant challenges, which require strong motivation and capabilities to overcome: there are too many reasons not to improve sanitation and not enough in favor. Households consistently aspire to a much higher-level solution than they can afford. Unable to afford what they want, they make do with what they have.

In Indonesia, the favored solution is a septic tank system, but most people are prepared to make do with a pour flush wet pit system. In Peru, people would like to have a bathroom with a toilet connected to the water network. Some make do with a "false toilet" with walls and a roof made from durable local materials even though there is no water supply. But even that is often out of the financial reach of poor families, who share their neighbors' toilets or use latrines. In Tanzania, many people interviewed would prefer a flush to pit latrine, but they recognize that they probably have to make do with a ventilated dry pit latrine with walls and (sometimes) a roof made from local material, such as maize stalks, jute bags, and sticks.

Nearly 170 million people in the four countries have unsatisfied sanitation aspirations. At least 90 million people living above the poverty line are "making do" with unimproved sanitation or sanitation below the standard they aspire to. If better-off families are prepared to make do, there is not much of an emulation push for poorer households to move up the sanitation ladder.

Lack of Commercial Viability and Inappropriate Business Models

Poor people want good-quality products that are simple to maintain, accessible service, credibility and choice, and complete service. Enterprises are providing them with discrete services, selling sanitation components, manufacturing components, and providing construction and pit-emptying services. Most of these activities are profitable, with some enterprises, particularly in Indonesia and Peru, having the potential to generate higher levels of margins through value-adding. But the industry is characterized by very localized microenterprises with low turnover and limited access to financial resources. As the prevailing technology is generic, and focused on manufacture by microenterprises, it does not lend itself to branding or coordinated marketing. Few enterprises invest in marketing to increase their sales. Even fewer have the business skills to realize how they might create more value.

Enhancing their ability to bundle services may be one way sanitation enterprises could exploit their "proprietary" capital—their knowledge of the market—and help reduce transaction costs for households. Although some enterprises are able to do so to a limited extent, few offer turnkey solutions. Many recognize that bundling and expanding the scope of their activities is important to their customers, but doing so, or pursuing more nuanced marketing activities, involves investment, which enterprises are reluctant to make.

Another way of increasing profitability would be to reduce costs, but enterprises have few options for doing so. With current technologies, inputs are dominated by materials whose prices are not within the control of sanitation enterprises. Production of two key materials, cement and steel, is dominated by a few companies in each country with localized monopoly power. Cement and steel account for about half the cost of production of a typical latrine or toilet set (slab plus three rings) in Bangladesh and 65 percent of the costs of making a slab in Tanzania. There is limited scope to reduce price, except by skimping on materials, with a consequent impact on durability and safety. Given their weight and volume to value ratios, distribution costs can be a significant part of sanitation costs to households in rural areas, where transport infrastructure is weak.

Fragmented and Uncoordinated Supply Chain

The most significant obstacle to scaled-up private provision of improved sanitation lies in the fact that the industry is not supplying products people want to buy. One factor preventing better alternatives from being offered is the fragmented supply chain, in which independent enterprises manufacture or supply one or more types of materials or pieces of equipment. For most manufacturers, importers, and retailers, sanitation represents a very small part of their overall sales. The availability of construction materials is thus driven by the demand for construction activities in other sectors.

Households typically help construct their latrines and toilets. Particularly where households do not have a latrine or toilet in their home, purchasing an improved sanitation solution can be challenging, because households often have to aggregate components and coordinate construction themselves. Enterprises make very little effort to market sanitation solutions or to improve coordination, exert quality control, or reduce costs within the supply chain. Actors that have the resources to address this challenge do not see sanitation as an important part of their market, and the enterprises closest to the market are very small and constrained in geographic reach. Few of these enterprises specialize in sanitation services, and they find it hard to signal any unique quality of service outside of the immediate vicinity where reputation is attested to by word of mouth.

Attitudes toward Investment and Serving the Poor

Given current demand, expanding coverage of improved sanitation among poor households will generally require expanding production capacity, relocating capacity to areas where demand exists, investing in marketing, bundling market offers, and developing and adopting new materials and technologies. Are enterprises moving in this direction?

Interviews reveal that enterprises in all countries recognize that the market for sanitation is growing, but they are concerned about the regularity of demand. A significant number of enterprises in Indonesia were planning to expand the range of sanitation-related services they offered, responding to signals from customers about their desire for service bundling. In contrast, in Bangladesh, enterprises contemplating investment were focused on expanding the scale of what they already do: manufacturing and selling latrine and toilet components. Few had any interest in expanding into installation, repair, or other sanitation-related business lines. The same attitude was evident in Tanzania.

Perceptions of the poor as an attractive customer segment vary. In Bangladesh and Indonesia, more than 60 percent of enterprises agreed or strongly agreed that the poor were target customers for them. This figure was just 48 percent in Tanzania, where a third of enterprises strongly disagreed that this was the case. More than three-quarters of Bangladeshi enterprises indicated that the poor do not pay on time, a view shared by smaller majorities in Indonesia (54 percent) and Tanzania (63 percent). More than three-quarters of enterprises in Tanzania indicated that the poor live in areas that are expensive to service because of transport and infrastructure problems.

Unsupportive Investment Climate

Broad government policies do not appear to be having much effect on surveyed enterprises, which are typically too small and too localized in reach to be affected by constraints that affect formal sector enterprises. There is little evidence that these enterprises are even aware of government sanitation policies and programs: more than 90 percent of enterprises in Bangladesh, 60 percent in Peru, and 40 percent in Tanzania either did not know about government policies or indicated that the policies had not been publicized in a way that helped them look out for business opportunities. Where governments have been involved in the direct supply of sanitation services to poor households, the top-down approach has not been very successful, but government provision and subsidies do not seem to be a significant source of distortion of the market.

Enterprises believe that governments should concentrate on removing risks to entry by providing market intelligence and promoting the entry of enterprises that are able to undertake transformative research and development on new technologies and materials. They believe that the poor quality and high cost of transport and the lack of adequate access to finance are obstacles to increased investment.

Recommendations

The study's recommendations focus primarily on the constraints inherent in current technologies and in the supply chains that support provision of on-site sanitation services. It is these constraints that lead to households being offered products and services that they are not very interested in buying. The recommendations are aimed at governments, development partners, and industry.

Stimulate Demand by the Poor

1. Enhance consumer awareness by improving household understanding of improved sanitation and complementing private marketing of sanitation

solutions to fill gaps in community understanding and address misinformation about the capabilities and maintenance requirements of improved on-site sanitation.

- Develop education and awareness programs that directly target households that already have some kind of sanitation to complement programs targeting open defecation, and address limited household understanding of the characteristics of improved sanitation systems.
- Ensure that campaigns address the gender dimensions of sanitation awareness and decision making where appropriate.

2. Improve affordability by smoothing and subsidizing sanitation expenditures to help very poor households mobilize cash to pay for improved latrines/toilets, using instruments that do not distort markets.

- Develop and support facilities that enable payment on installment terms, either intermediated through agency arrangements with manufacturers and suppliers of components or through financial institutions that provide consumer loans to households.
- Develop and finance targeted subsidies for extremely poor households in locations where suitable technology cannot be delivered at reasonable costs.

Encourage Innovation and Facilitate Efforts to Relax Business Model and Supply Chain Constraints

3. Spur innovation by stimulating (and if necessary financially supporting) the development of affordable technologies with consumer appeal.

- Help develop technologies (preferably proprietary or licensable) that use materials that are light and easy to transport; easy to clean; and amenable to mass production, branding, and marketing through distribution networks coordinated and supported by manufacturers.
- Assist in the development of modular technologies that facilitate incremental improvements to sanitation facilities as household interest grows and households are able to mobilize funds.
- Explore options for stimulating research and development by the private sector (for example, through patents, contracts, and grants).
- If the preferred model of commercial development and rollout of proprietary technology is not forthcoming, consider expanding funding by the international development community for research and development to develop technologies that are appropriate for delivery through a market-based system.

4. Encourage larger businesses to enter the on-site sanitation sector by fostering entry of well-capitalized enterprises with marketing skills to drive consumer interest, and capacity to coordinate supply chains and support installation and maintenance by small-scale local enterprises.

- Support the collection and dissemination of market intelligence, such as information on the size and nature of the market, including the fact that it includes many households that are above the poverty line.

- Explore options for incentives to entry, including start-up financing and support.
- Encourage the formation of associations of enterprises involved in sanitation to develop a distribution channel to the "last mile" and to assist in the dissemination of market and technical information.

5. Enable quality assurance and accreditation. With the entry of larger businesses in the supply chain, assist microenterprises at the front end to more credibly signal service quality to a larger market and assure potential purchasers that they will get value for money and durability and continuity of service.
   - If capacity exists, introduce public sector certification of technologies or government endorsement of international certification by development partners, but avoid government regulation of standards.
   - Facilitate industry-based accreditation systems for enterprises or solutions to enable manufacturers to offer warranties on installation.

6. Support business capacity development by helping the microenterprises currently delivering the bulk of on-site solutions to expand their limited business expertise so that they can better participate in an expansion of supply.
   - Facilitate capacity building through partnerships with larger actors in the supply chain in agency, distribution, or subcontracting networks that also address the capacity and commercial challenges at the front end of the supply chain.
   - Develop elements of public sector sanitation marketing and education campaigns that can be used by small-scale providers of private sanitation services.

Improve the Investment Climate and Sectoral Policies

7. Facilitate private provision by clearly spelling out an active (rather than default) role for the private sector in government strategies and policies, and improve sector investment planning to identify markets with potential for private participation.
   - Detail and publicize policies to facilitate the private sector role. Identify and resource responsibilities across different levels of government for implementation, especially where local governments have responsibility, mandates, and resources for sanitation.

8. Regulate septage disposal by formulating practical standards and protocols for disposal of fecal sludge and by building capacity to implement them, in order to develop safe arrangements for disposal to accompany the growth of private sector pit and septic tank emptying.
   - Develop treatment sites and protocols for treatment.
   - Explore options for financing disposal sites, including public-private partnerships.

CHAPTER 10

What Is the Problem?

In many developing countries, significant numbers of poor and nonpoor households do not use improved sanitation—a facility that hygienically separates human excreta from human contact. (Appendix table B.1 describes the various types of improved and unimproved sanitation.) Lack of access is more common among the poor, however, and poor people are less equipped to deal with the personal and economic consequences of poor sanitation. Illness leading to loss of productivity of income earners can have a catastrophic effect on poor households, which may also be less able to afford treatment.

Access Is Inadequate

Despite substantial increases over the past two decades, access to improved sanitation remains limited in the case study countries (Bangladesh, Indonesia, Peru, and Tanzania) (figure 10.1a). It is particularly low in rural areas (figure 10.1b). The nature of the challenge of improving access differs across the four countries. It does not appear to be directly correlated with the level of economic development. The proportion of the rural population still resorting to open defecation is much lower in rural Tanzania (16 percent), for example, than in countries with much higher average levels of income, such as Indonesia (36 percent) and Peru (28 percent) (figure 10.2).[1]

Poor Sanitation Imposes Very High Costs on Developing Countries

Poor sanitation imposes very high costs on developing countries. In the four countries covered by this study, the total economic losses have been estimated to be well over $10 billion a year, an astonishing 1.0–6.3 percent of each country's gross domestic product (GDP) (table 10.1).

Figure 10.1 Access to Improved Sanitation in Bangladesh, Indonesia, Peru, and Tanzania, 1990 and 2010

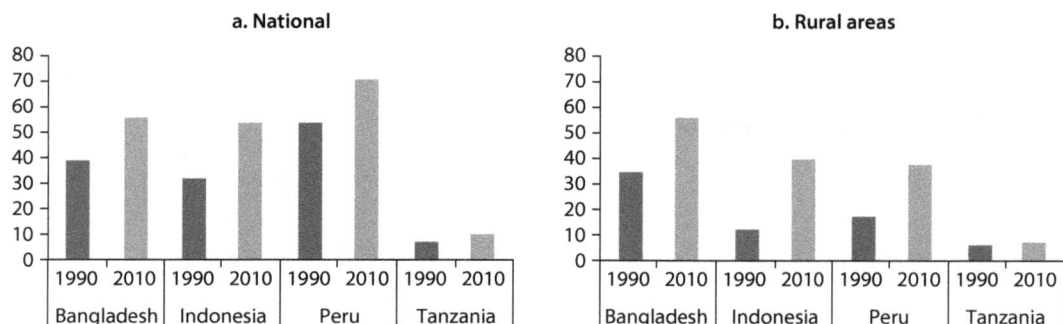

a. National

b. Rural areas

Sources: WHO/UNICEF 2013a, 2013b, 2013c, 2013d.
Note: Unless otherwise indicated, data for tables and figures come from the country case studies (see references).

Figure 10.2 Types of Sanitation Used in Bangladesh, Indonesia, Peru, and Tanzania, 2010

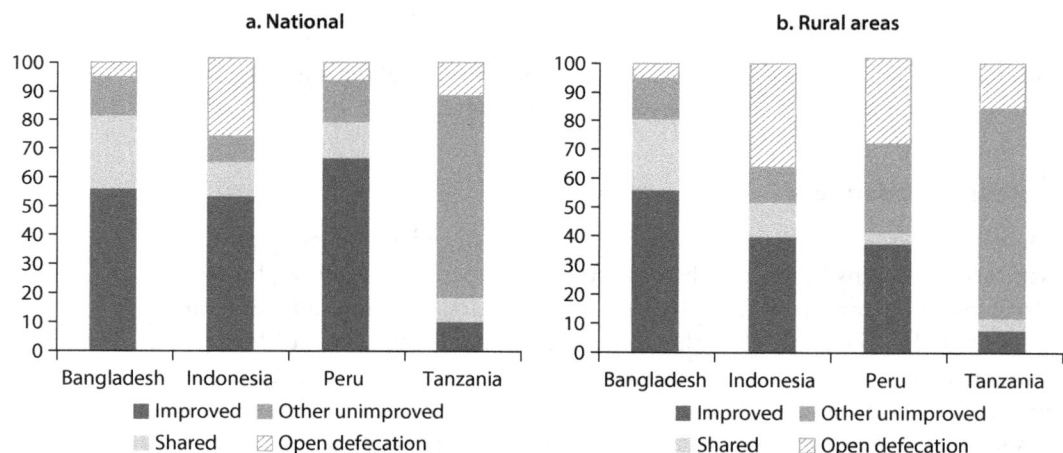

a. National

b. Rural areas

■ Improved   ■ Other unimproved

▨ Shared   ▨ Open defecation

Sources: WHO/UNICEF 2013a, 2013b, 2013c, 2013d.

Table 10.1 Costs of Inadequate Sanitation in Bangladesh, Indonesia, Peru, and Tanzania

Country	Cost (millions of dollars)	Percentage of GDP	Cost per capita (millions of dollars)
Bangladesh	4,200	6.3	28
Indonesia	6,300	2.3	27
Peru[a]	759	1.1	27
Tanzania	206	1.0	5

Sources: Larsen and Strukova 2006; WSP 2013.
a. Includes costs of inadequate water supply.

Governments Cannot Solve the Problem

Private enterprises may be underinvesting in the sanitation sector because the social benefits of improved sanitation are not reflected in the price that consumers are willing to pay. The existence of these "externalities" does not necessarily provide a rationale for government provision of sanitation, however. Moreover, even if it did, in most developing countries with large numbers of poor people, the government lacks the financial and organizational capacity to meet the need for improved sanitation from public resources.

In the countries covered by this study, most poor (and many nonpoor) households look to the private sector to help meet their sanitation needs. In Bangladesh, Indonesia, and Tanzania, at least 95 percent of the population with toilets rely on private initiatives to construct their facilities. In highly urbanized Peru, with a strong tradition of public utilities providing sewerage systems, a quarter of people with some kind of sanitation use privately constructed septic tanks and latrines/toilets.

Note

1. In 2011, on a purchasing power parity basis, per capita gross national income (GNI) was $1,940 in Bangladesh, $4,500 in Indonesia, $9,440 in Peru, and $1,500 in Tanzania, according to the World Development Indicators.

References

Country Studies

Akademika. 2013. *Global Study for the Expansion of Domestic Private Sector Participation in the Water and Sanitation Market.* Jakarta: Akademika.

DevCon (DevConsultants Limited). 2013. *Sanitation Bangladesh: Global Study for the Expansion of Domestic Private Sector Participation in the Water and Sanitation Market.* Dhaka: DevCon.

IMASEN and Ausejo Consulting. 2013. *Peru Country Report: Global Study for the Expansion of Domestic Private Sector Participation in the Water and Sanitation Market.* Lima: IMASEN and Ausejo Consulting.

PATH. 2012. *Market Assessment of Domestic Private Sector Provision of Household Sanitation in Tanzania, Final Country Report.* Seattle and Washington, DC: PATH.

Other References

Larsen, B., and E. Strukova. 2006. "Peru: A Cost-Benefit Analysis of Improved Water Supply, Sanitation and Hygiene and Indoor Air Pollution Control." Background paper prepared for the World Bank report *Republic of Peru Environmental Sustainability: A Key to Poverty Reduction in Peru.* Washington, DC: World Bank. http://documents.worldbank .org/curated/en/2007/06/7910058/republic-peru-environmental-sustainability-key -poverty-reduction-peru.

WHO (World Health Organization)/UNICEF (United Nations Children's Fund) Joint Monitoring Program. 2012. "Types of Drinking-Water Sources and Sanitation." http://www.wssinfo.org/definitions-methods/watsan-categories/.

————. 2013a. "Estimates for the Use of Improved Sanitation, Bangladesh." http://www
.wssinfo.org/fileadmin/user_upload/resources/BGD.xlsm.

————. 2013b. "Estimates for the Use of Improved Sanitation, Indonesia." http://www
.wssinfo.org/fileadmin/user_upload/resources/IND.xlsm.

————. 2013c. "Estimates for the Use of Improved Sanitation, Peru." http://www.wssinfo
.org/fileadmin/user_upload/resources/PER.xlsm.

————. 2013d. "Estimates for the Use of Improved Sanitation, Tanzania." http://www
.wssinfo.org/fileadmin/user_upload/resources/TZA.xlsm.

WSP (Water and Sanitation Program). 2013. "Economics of Sanitation Initiative."
Washington, DC: World Bank. http://www.wsp.org/content/economic-impacts
-sanitation.

CHAPTER 11

Why This Study?

In developing countries, publicly supplied sanitation services fail to reach most poor (and many not-so-poor) people. In recent years, attention has focused on the contribution of the domestic private sector and market-driven solutions to expand the use of improved sanitation. Governments have taken various approaches. In some countries, they have left sanitation almost entirely to the private sector and households. In others, they have only recently recognized private provision in their national sanitation strategies and begun exploring ways to facilitate an expanded role for the domestic private sector.

This study examines the involvement of the domestic private sector in the construction of on-site sanitation facilities and the delivery of sanitation services in rural areas and small semi-urban settlements. Its aim is to understand the extent to which private sector schemes can provide the poor with improved sanitation.

This study considers two sets of factors—commercial factors and investment climate factors—that affect enterprises' actual or perceived costs and risks and, in turn, their decisions to invest in the provision of on-site sanitation services (figure 11.1). It examines both sets of factors by seeking answers to the following questions:

- Is lack of interest by the domestic private sector a rational response to weak market potential, or are lack of enterprise viability and the use of inappropriate business models preventing it from taking advantage of market opportunities?
- Are investment climate factors increasing the (actual or perceived) costs and risks associated with doing business?

To shed light on these issues, the study team conducted research into the sanitation sector and its policy environment, surveyed suppliers of on-site sanitation facilities and services, held focus group discussions with actual and potential customers of these suppliers, and interviewed other stakeholders, including government officials, in Bangladesh, Indonesia, Peru, and Tanzania.[1] The country

Figure 11.1 Study Analytical Framework

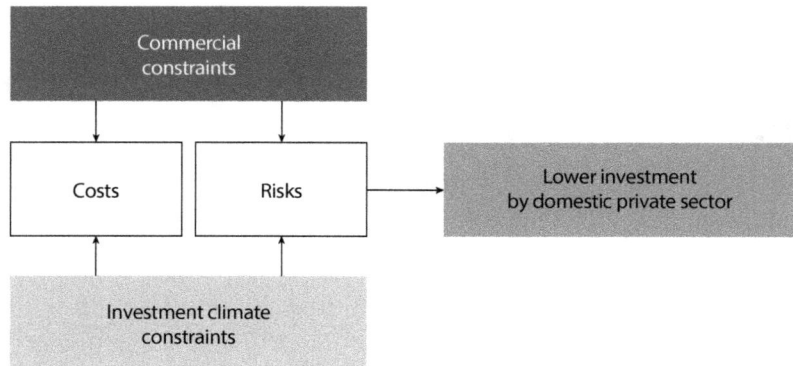

studies focused on rural areas and small semi-urban settlements. In the four countries, a total of 109 enterprises were surveyed, and focus group discussions were held with 682 people from poor households. The study teams also consulted with enterprises involved in the supply chain that were not directly providing services to poor households and with officials and staff from relevant government and nongovernment agencies.

Note

1. The data from surveyed enterprises in Peru were more limited than they were for the other countries.

CHAPTER 12

On-Site Sanitation Services in the Case Study Countries

The country case studies focused on on-site sanitation services, where the private sector plays a large role.[1] They looked at a range of private enterprises providing on-site sanitation services, including enterprises manufacturing and selling latrine and toilet components, building sanitation facilities, and providing emptying and disposal services (table 12.1).

In Bangladesh, Indonesia, and Tanzania, all rural people and the majority of people in urban areas use on-site sanitation (pit latrines and septic tank systems). In Peru, which is much more highly urbanized, nearly two-thirds of the population has access to a sewer network, including 12 percent of the rural population. For the rural poor, however, on-site facilities are the only type of improved sanitation.

Table 12.1 Type and Location of Sanitation Enterprises Interviewed for Country Case Studies

Country	Type of enterprise	Site and reach	Number of enterprises
Bangladesh	Prefabricated concrete producers casting cement platforms and rings and constructing latrines	Rural villages in eight subdistricts	30
Indonesia	Producers of sanitation facilities, including toilets and septic tanks	Secondary towns in seven districts	22
	Truck-based septic tank emptying and disposal companies	Three cities	10
Peru	Regional component suppliers and hardware stores	Three cities and two towns	7
	Construction and plumbing service companies (including an association of plumbers)	One city	4
	Water and sanitation operators	One town	2
	Regional water and sanitation utilities (public providers that operate sewer systems)	Three cities and one town	4
Tanzania	Masons involved in casting sanitation slabs and installation	Rural villages in three districts	9
	Hardware stores selling components and casting sanitation slabs	Two districts	21

Table 12.2 Improved Sanitation Options Available to Poor Households in Bangladesh, Indonesia, Peru, and Tanzania

Type of facility	Bangladesh	Indonesia	Peru	Tanzania
Above ground	Water-sealed pour flush pan on concrete slab	Water-sealed ceramic pour flush pan on concrete slab	Water-sealed pour flush pan on concrete slab	Concrete slab on wooden floor
Below ground	Pit lined with three concrete rings	Concrete-lined pit	Concrete-lined pit	Unlined pit
Superstructure	Bamboo housing with plastic roof	Brick housing	Drywall housing	Local materials
Collection and disposal	Manual pit emptying and burying by households or paid labor	Pit emptying by vacuum trucks and disposal into sludge treatment facilities	On-site disposal	Closing pit off when full and moving

The type of on-site facility used varies across countries. The case studies focused on a set of options that are typical in poor rural areas (table 12.2).

Note

1. In Peru, where 12 percent of the rural population and 66 percent of the total population have access to sewerage, the study authors also interviewed sewer operators, to provide context.

CHAPTER 13

Is Market Potential Sufficient to Justify Private Investment?

Between 2000 and 2010, 15 million households in Bangladesh, Indonesia, Peru, and Tanzania acquired improved sanitation facilities for the first time. Supplying these households with the kinds of sanitation options currently marketed in each country cost an estimated $800 million, or $80 million a year (table 13.1). Over the same period, the entire stock of latrines/toilets and septic tanks in place in 2000 probably needed replacing, at an average annual cost of about $220 million a year. Putting the two figures together yields an estimated size of the (rural and urban) sanitation sector in the four case study countries of $300 million a year.

How large can the market become? The Joint Monitoring Program of the World Health Organization and United Nations Children's Fund (UNICEF) estimates that about 228 million people in the four countries lack access to improved sanitation. Meeting the needs of these people would involve sales of about $2.6 billion (table 13.2).[1] About 70 percent of these households are in rural areas. Less than a third of them live below the national poverty line (poor people account for about $700 million of this market). Once these people are served, the market for providing them with replacement equipment would be worth about $550 million a year.

There is also a potentially significant market for repairing latrines and toilets and emptying and disposing of septage. In Bangladesh, about three-quarters of latrines do not have a functioning water seal. In Indonesia, some 37 million households have pits or septic tanks that need periodic emptying: a conservative estimate suggests that the potential market for truck-based emptying services there is about $100 million a year.

Economic Drivers

Real per capita incomes have been rising in all four countries, and the proportion of the population living below the poverty line has been falling (table 13.3). Both figures suggest that the aggregate ability to pay for improved sanitation should be increasing.

Table 13.1 Estimated Sales of New and Replacement Improved Sanitation in Bangladesh, Indonesia, Peru, and Tanzania, 2000–10

Item	Bangladesh	Indonesia	Peru	Tanzania	Total
Extension of service to new customers					
Number of households (millions)	5.0	8.7	0.9	0.3	14.9
Sales (millions of dollars)	151	556	79	8.0	795
Replacement of facilities by existing customers					
Facilities needing replacement (million)	13.9	22.9	3.2	0.6	40.6
Sales (millions of dollars)	416	1,466	297	18	2,197

Sources: Estimates of improved sanitation coverage and population are from WHO/UNICEF 2013a, 2013b, 2013c, 2013d; 2012 costs of commonly used improved sanitation facilities in each country are from country studies. Unless otherwise indicated, data for tables and figures come from the country case studies.

Table 13.2 Estimated Potential Expansion of Market for Improved Sanitation in Bangladesh, Indonesia, Peru, and Tanzania

Estimate	Bangladesh	Indonesia	Peru	Tanzania	Total
Size of market (millions of people not using improved sanitation in 2010)[a]					
Whole country	66.2	111.4	8.5	41.6	227.8
Rural areas	48.4	81.8	4.2	31.5	165.9
Urban areas	17.9	29.6	4.3	10.2	69.2
Poor[b]	54.2	12.5	3.5	12.6	97.4
Nonpoor	12.1	98.9	5.0	29.0	148.4
Market value (millions of dollars)					
Whole country	452	1,739	155	240	2,587
Rural areas	330	1,278	77	182	1,866
Urban areas	122	462	79	59	720
Poor[b]	369	196	64	73	701
Nonpoor	82	1,543	92	167	1,885

Sources: WHO/UNICEF 2013a, 2013b, 2013c, 2013d reports; country studies.
a. Figures reflect the Joint Monitoring Program definitions of improved sanitation (see appendix table B.1). Bangladesh uses a slightly different definition; it includes conforming latrines that are shared by a maximum of two households as improved. Based on this definition, the number of people not using improved sanitation is 28.1 million.
b. Defined using national poverty line.

Urbanization will also affect the kinds of sanitation solutions the market will require. In Bangladesh, the absolute number of people living in rural areas is projected to start declining by around 2020, as migration to towns and cities continues and urbanization of rural areas accelerates. In Tanzania, where sewered systems are extremely limited, the urban population is growing nearly 70 percent faster than the population as a whole.

Policy Drivers

The impact of sector policies in developing the sanitation market is limited and has not promoted increased participation from the private sector. Public policies have tended to focus on infrastructure investment rather than setting

Table 13.3 Per Capita Gross Domestic Product and Poverty Headcount in Bangladesh, Indonesia, Peru, and Tanzania, 2000 and 2010

Indicator	Bangladesh	Indonesia	Peru	Tanzania
Per capita GDP (2005 U.S. dollars)				
2000	970	2,623	5,547	868
2010	1,488	3,885	8,555	1,293
Change (percent)	53	48	54	49
Poverty headcount				
2000 (percent)	49	19	48	36
2010 (percent)	32	13	28	33[a]
Change (percentage points)	−17	−6	−20	−3

Sources: World Development Indicators 2013 and country studies.
a. Data are for 2007.

a framework for market provision of services. Recent policy statements have begun to emphasize the role of government in creating demand for sanitation services and should have a positive effect in the future. But the lack of articulation of the role for the private sector and how it might be facilitated could frustrate this intent.

Current policies have not promoted private sector participation, but they do not seem to have hindered it (table 13.4). Focus group discussions reveal that rural households, poor and nonpoor, believe that sanitation ought to be a publicly provided service, but they recognize that they will have to look after their own needs.

Interviews with service providers indicated that policies and government agencies are seen as largely irrelevant to their business. When asked for an opinion about the clarity of rules and standards for sanitation, nearly all surveyed enterprises in Bangladesh said they did not know. In Peru and Tanzania, about half of the surveyed enterprises disagreed or strongly disagreed that rules were clear. Asked whether sanitation promotion programs were well publicized so that enterprises can look out for business opportunities, a similar pattern of responses emerged: more than 90 percent of enterprises in Bangladesh did not know, and 60 percent of enterprises in Peru and 40 percent in Tanzania disagreed or strongly disagreed. Only in Indonesia did a large majority of enterprises think that the rules were clear (80 percent) and provided opportunities to look out for business (90 percent). These figures probably reflect the fact that the government has made it clear that sanitation is a private responsibility.

Rethinking Market Drivers

Beyond broad market drivers lies the complexity and diversity of household preferences and aspirations. Both affect the value households place on sanitation and the expectations they have with respect to how sanitation solutions suit their needs.

Table 13.4 Policy Drivers of Sanitation in Bangladesh, Indonesia, Peru, and Tanzania

Item	Bangladesh	Indonesia	Peru	Tanzania
Policy goal				
Private sector engagement for on-site sanitation	No evidence of specific policy.	Regulation is in place governing public-private partnerships for private operation of sanitation infrastructure where users cannot opt out.	No evidence of specific policy. Recent policy statements express desire to engage private sector.	Water Act of 1966 includes policy statement on public-private partnership but no implementation regulations.
Subsidy to households for on-site sanitation	Directive for local governments to allocate 20 percent of development budget to support poor households' access to sanitation.	De facto policy of no subsidy, but programs have invested in communal and household-level sanitation facilities.	Policy of no subsidy for household investment in sanitation, including connecting to sewers. Recent statements emphasize need for providing low-cost alternatives to sewers.	Policy of no subsidy for sanitation.
Promotion of sanitation	Traditionally, strong public focus on hygiene and sanitation promotion implemented through localized, subdistrict committees.	Recent policy emphasis on demand creation role of government.	Not clear.	Traditionally, strong public focus on sanitation. National program launched in 2010 for promotion and marketing of sanitation.
Regulation of on-site sanitation	Local committees emphasize promotion rather than enforcement or regulation.	Standards exist, but implementation varies across local governments.	Approaches only just being developed for on-site solutions for rural and poor households. Traditionally focused on sewers.	Local institutions monitor that households have a latrine.
Operationalization				
Public investment programs	Free latrines distributed.	Cost-sharing of rural household sanitation, communal facilities, and investment in municipal facilities.	Public sector finances development of network water and sanitation. Some municipalities provide free toilets.	Focus is on school facilities.
Instruments for delivering sanitation programs	Local government institutions and nongovernmental organizations.	Community and public health institutions.	State utilities and community water operators.	Line ministries and local government institutions.
Market-based approaches supported by public programs	To be tested for first time at scale under national program.	Proof of concept stage; scaled testing underway.	Proof of concept stage; some targeted pilots being conducted.	Proof of concept stage; scaled testing underway.

Household income is not always a reliable predictor of demand for improved sanitation. In Indonesia, according to the national economic and social survey, more than 9 million rural households who resort to open defecation are nonpoor—this number represents more than 20 percent of all rural households in the country. In Peru, nearly three-quarters of nonpoor households living within the sewer network choose not to connect to it. In Tanzania, nearly 17 million nonpoor rural people (about 85 percent of the rural nonpoor population) use unimproved sanitation—about 400,000 households that are wealthy enough to have cement floors in their houses do not have a slab in their latrines that would meet the standards of improved sanitation (figure 13.1).

Focus group discussions reveal that households look for qualities in their facilities without reference to what government or international standards may define as "improved sanitation." In Bangladesh, there also seems to be a willingness to share latrines not observed in the other countries. Poor households in Bangladesh indicated preferences for different features of an improved latrine: more than 20 percent indicated a preference for a raised platform to provide safety from floods, and a similar proportion opted for a superstructure with bamboo walls and corrugated iron roofing. In Peru, households at all income levels that have a regular supply of water (80 percent of all households) regarded latrines as a symbol of poverty and social exclusion. They aspired to a bathroom with a sink and a shower or at the least a "false toilet" with all the appurtenances of a bathroom except the connection to a sewerage network. Interest in improved sanitation is very low in Tanzania, even though—or, perhaps,

Figure 13.1 Opportunities for Providing Improved Sanitation in Tanzania

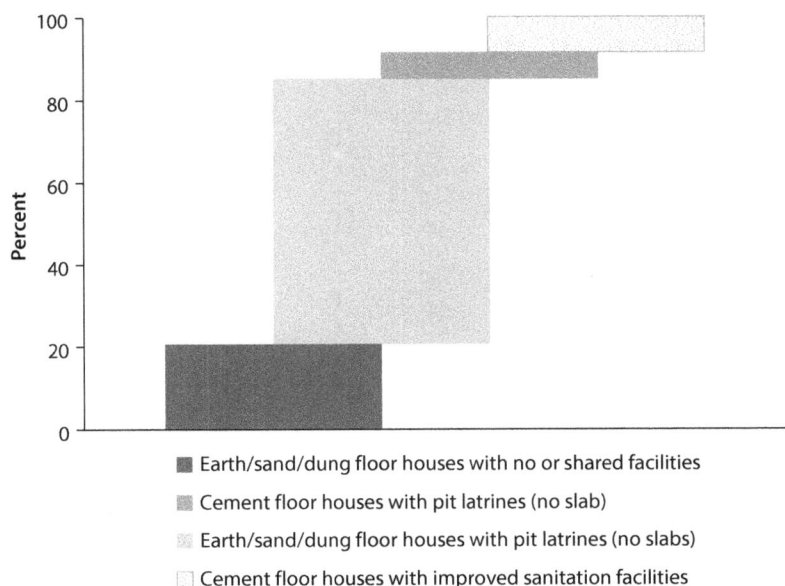

- Earth/sand/dung floor houses with no or shared facilities
- Cement floor houses with pit latrines (no slab)
- Earth/sand/dung floor houses with pit latrines (no slabs)
- Cement floor houses with improved sanitation facilities

Source: PATH 2012.

because—strong government programs after independence led to a very high coverage of basic sanitation.

Based on evidence on people's preferences, it appears that nearly 170 million people in the four case study countries have unsatisfied aspirations or wants with respect to improved sanitation. Between 60 percent (Bangladesh and Indonesia) and 100 (Tanzania) percent of rural poor people fall into this category (figure 13.2). Large numbers of nonpoor rural people—more than 60 million in Indonesia and 90 million in total—also have unsatisfied aspirations.

Figure 13.2 Unsatisfied Sanitation Aspirations of Poor Households in Bangladesh, Indonesia, Peru, and Tanzania, 2012

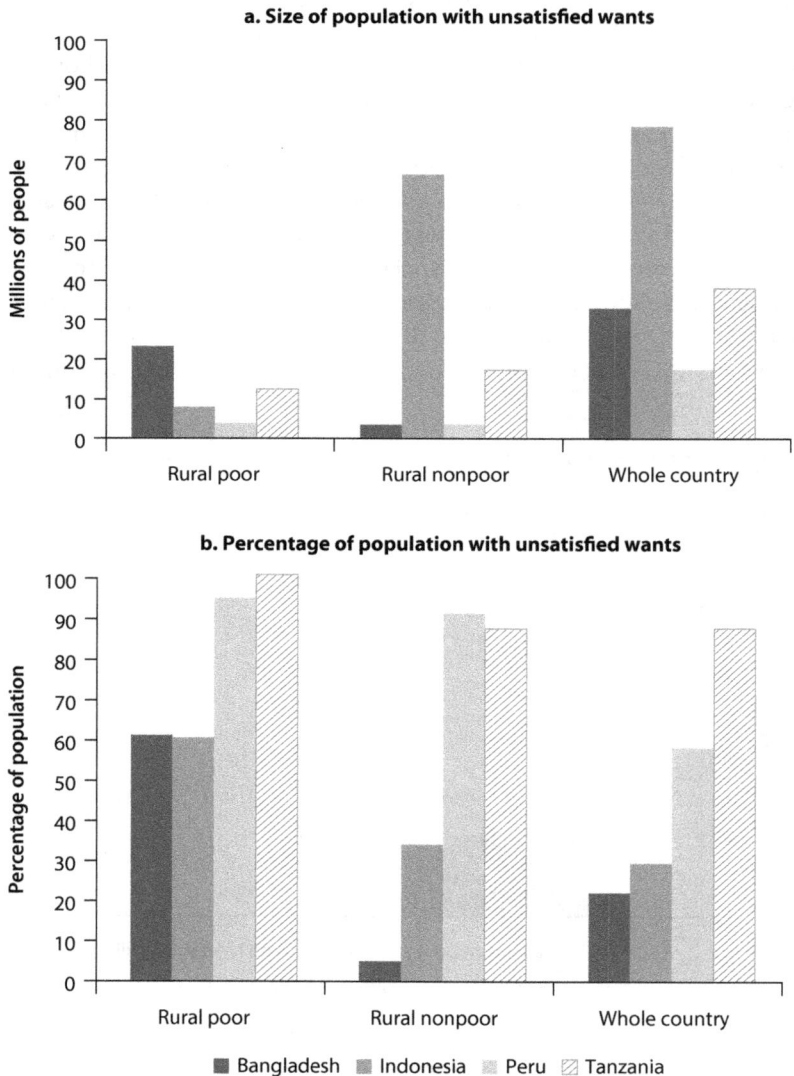

Note: Poverty defined using national poverty lines.

The benefits of moving from open defecation to basic sanitation are much larger than the benefits of moving to the next stage of improved sanitation, and the incremental costs of moving up the ladder are relatively high.[2] The gap between what households want/expect in their sanitation solutions and what solutions are most cost-effective in delivering benefits has important implications for public programs seeking to adopt market-based approaches.

Notes

1. For information on the Joint Monitoring Program, see WHO/UNICEF 2012.

2. For example, a 2011 study by the Water and Sanitation Program shows that the cost-benefit ratio for moving from open defecation to a shared latrine in a rural district in Indonesia was 5.4, whereas the ratio for moving from a shared latrine to a private septic tank was 2.0.

References

Country Studies

Akademika. 2013. *Global Study for the Expansion of Domestic Private Sector Participation in the Water and Sanitation Market.* Jakarta: Akademika.

DevCon (DevConsultants Limited). 2013. *Sanitation Bangladesh: Global Study for the Expansion of Domestic Private Sector Participation in the Water and Sanitation Market.* Dhaka: DevCon.

IMASEN and Ausejo Consulting. 2013. *Peru Country Report: Global Study for the Expansion of Domestic Private Sector Participation in the Water and Sanitation Market.* Lima: IMASEN and Ausejo Consulting.

PATH. 2012. *Market Assessment of Domestic Private Sector Provision of Household Sanitation in Tanzania, Final Country Report.* Seattle and Washington, DC: PATH.

Other References

WHO (World Health Organization)/UNICEF (United Nations Children's Fund) Joint Monitoring Program. 2012. "Types of Drinking-Water Sources and Sanitation." http://www.wssinfo.org/definitions-methods/watsan-categories/.

———. 2013a. "Estimates for the Use of Improved Sanitation, Bangladesh." http://www.wssinfo.org/fileadmin/user_upload/resources/BGD.xlsm.

———. 2013b. "Estimates for the Use of Improved Sanitation, Indonesia." http://www.wssinfo.org/fileadmin/user_upload/resources/IND.xlsm.

———. 2013c. "Estimates for the Use of Improved Sanitation, Peru." http://www.wssinfo.org/fileadmin/user_upload/resources/PER.xlsm.

———. 2013d. "Estimates for the Use of Improved Sanitation, Tanzania." http://www.wssinfo.org/fileadmin/user_upload/resources/TZA.xlsm.

World Development Indicators (database). 2013. Washington, DC: World Bank. http://data.worldbank.org/data-catalog/world-development-indicators.

CHAPTER 14

What Affects Demand for On-Site Sanitation?

Affordability is an important determinant of demand for on-site sanitation. The fact that many poor households without sanitation own mobile phones suggests that the poor are willing to pay for value and that affordability is not the only factor, however. The more important constraint on increasing access is the low value people place on the improved sanitation options available in the market.

Cost

The on-site sanitation solutions offered to poor households by the private sector are similar in the four countries surveyed, partly because of efforts made by the international community to develop and promote a set of inexpensive and easy-to-produce solutions. Except in Peru, these solutions typically cost 3–4 percent of the annual income of households living below the poverty line—not an insurmountable cost barrier (table 14.1).

Cash Constraints

The poor households interviewed owned many consumer durables, such as mobile phones, motorbikes, and bicycles. Many of these households spend as much on mobile phone use annually as it would cost to install the available sanitation solution. In Bangladesh, all of the poor households and 34 percent of the extremely poor households that participated in the focus groups had at least one mobile phone.[1] On average, households with a phone were spending $55 a year on the service, nearly twice the one-time cost of a standard improved latrine or toilet. The prevalence of mobile phone ownership among poor households suggests that households can make significant outlays for a valued service if expenditure can be spread out over time.

Purchasing improved sanitation requires a large outlay of cash at one time. Many poor households interviewed have uncertain and seasonally varying incomes; in many cases, a significant part of their consumption is also

Table 14.1 Estimated Costs of Toilets and Pit Emptying in Bangladesh, Indonesia, Peru, and Tanzania, 2012

Country	Country/on-site sanitation option	Cost per unit/ service (dollars)	Share of poor household monthly income (percent)	Share of poor household annual income (percent)
Bangladesh	Toilet	30	39	3
	Pit emptying	5	7	1
Indonesia	Toilet	64	44	4
	Pit emptying	15	11	1
Peru	Toilet	93	89	7
Tanzania	Latrine	30	48	4

Figure 14.1 Price Households in Bangladesh, Indonesia, Peru, and Tanzania Would Pay for Ideal Sanitation Facility

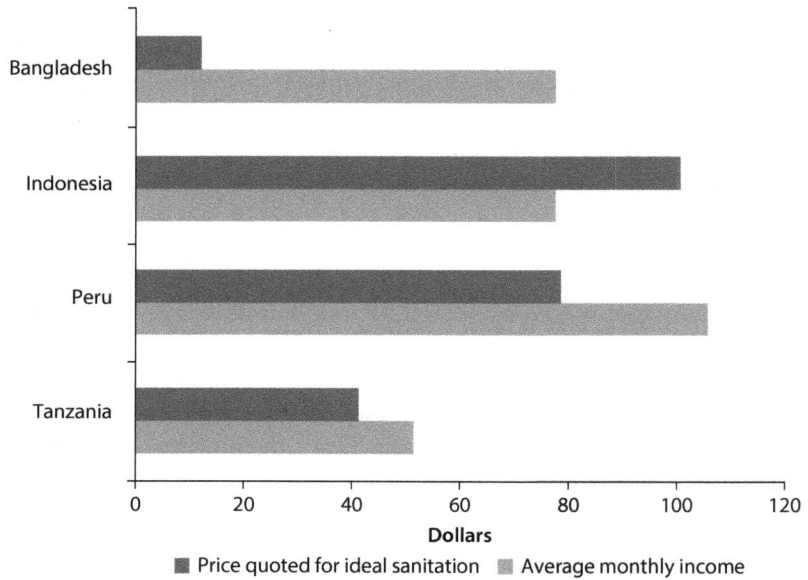

self-produced, so that cash income is less than total income. Indeed, when estimating what they thought they would have to pay for their ideal sanitation facility, focus group participants in all countries except Bangladesh generally cited a value that represented a significant portion of their average monthly household income (figure 14.1). Respondents were able to correctly identify the cost of ideal options. These figures suggest that poor people understand the order of magnitude of outlay involved in purchasing the kind of sanitation to which they aspire.

In all four countries, households strongly indicated that installment payment arrangements would enhance their willingness to spend on sanitation infrastructure. In Bangladesh, provision of an installment option was the service most frequently cited when households were asked what additional service they wanted

Figure 14.2 Additional Services Households in Bangladesh Would Like from Their Sanitation Providers

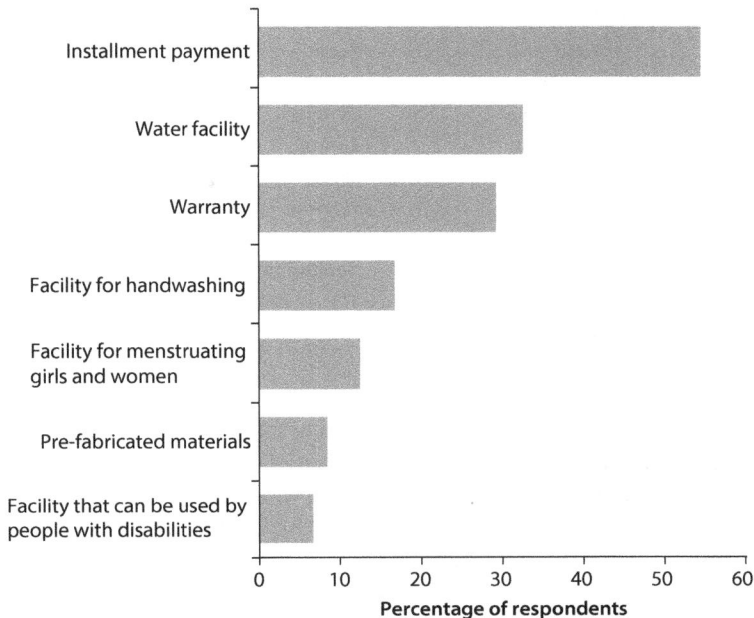

Source: DevCon 2013.

from providers (figure 14.2); in Indonesia, this feature was the second-most important consideration (after availability of land).

Importance of Sanitation to Households

To market on-site sanitation to the poor, private enterprises need to determine whether their products or services are the ones households want and how much the poor are prepared to pay for them. Focus group discussions with poor people revealed that improved sanitation is not an explicit or revealed priority for most of them (table 14.2).

In Indonesia, sanitation was not well understood even by village chiefs, who associated the term *sanitation* with a broader concept of cleanliness and garbage collection. In Tanzania, sanitation was not a priority among community members, who considered it a government responsibility. In both countries, sanitation was a low priority even where money was not an issue. In contrast, in Bangladesh, a large proportion of focus group participants had private latrines (69 percent) or were sharing latrines with other households (25 percent), and three-quarters of villages were engaged in a sanitation program sponsored by the government or a nongovernmental organization.

Although poor people are concerned about cost and expenditure, more than a lack of money seems to underpin the low priority they assign to

Table 14.2 Spending Priorities of Poor Households in Bangladesh, Indonesia, and Tanzania

Priority	Bangladesh	Indonesia	Tanzania
1	Food	Food	Health
2	Clothing	Health	Water
3	Education	Clothing	Education
4	Housing	Education	Housing
5	Health	Water	Furniture
6	Cell phones	Housing	**Sanitation**
7	Electricity	Communication	
8	Furniture	**Sanitation**	
9	Water	Transportation	
10	**Sanitation**	Recreation	

improving sanitation. The focus group discussions reveal that the poor faced limited options and significant challenges, which require enormous motivation and capabilities to overcome. There are too many reasons not to improve sanitation and not enough reasons to do so. As the Tanzania country study report points out:

> One of the most basic factors affecting the demand for domestic private sector provision of household sanitation is the level of priority households place on solving their sanitation solutions, especially as it relates to their ability and willingness to pay for such goods or services. For families that have either been defecating in the open or utilizing basic materials sourced freely from the surrounding environment, the idea of paying for sanitation would seem to present an extremely formidable mental obstacle—especially for a household with already constrained expendable income. Additionally, low-income households, particularly in rural areas, must deal with fluctuating and sometimes irregular inflows of money. For largely agricultural societies such as Tanzania, these fluctuations often ebb and flow with harvest seasons, but other seasonality considerations include school fees, holidays, and family events such as funerals and weddings. It is critical to better understand how households view sanitation as a priority and what might be done to help them consider its importance economically and within the context of other major needs within the family.

Acceptance of "Make Do" Solutions

In all four countries, the available solutions that are affordable were designed primarily to address key public health concerns using materials that are readily available throughout most developing countries. They may not be sufficiently attractive for households to acquire, however. Bangladesh uses a latrine design introduced in the 1970s to reduce the spread of waterborne diseases. In the intervening 40 years, there has been little product development to address households' needs or match their growing aspirations.

The focus group discussions conducted in all four countries probed what poor households would like to have in a sanitation solution and their ability and willingness to pay for it. Participants consistently revealed that they aspire to a solution at a level much higher than they can afford; sensing the futility of their desire, they "made do" with a less desired option. In Bangladesh, for example, coverage of wet pits is high, but many facilities do not function well (broken water seal). People who shared a facility were keenly aware of the burden they imposed on their neighbors. In Indonesia, most people would have liked to have had a septic tank system but were prepared to make do with a pour flush pit system. In Peru, respondents ideally wanted a bathroom with a toilet connected to a water network. Some people made do with a "false toilet," the walls and roof of which were made from durable materials even though there was no water supply. But even that was often out of the financial reach of poor families, who therefore shared their neighbors' toilets or used a latrine. In Tanzania, many respondents preferred a flush toilet to a pit latrine but recognized that they probably had to make do with a ventilated dry pit latrine with walls and (sometimes) a roof made from local materials, such as maize stalks, jute bags, and sticks.

These "make do" solutions leave many people unsatisfied with their current systems. People defecating in the open were concerned about the inconvenience of going outside at night and the risk of physical harm (from snake bites, for example). People using latrines complained of odors, the rapid filling up of the pit, the maintenance involved, and the fact that latrines are a temporary solution. For example, unimproved latrines are widely used in Peru, even though they do not provide great advantages and cause problems. Improved latrines are used only by some members of the family, particularly children; they are considered provisional and difficult to relocate. For this reason, some respondents indicated that they prefer to relieve themselves outside, in order to keep their latrines from filling up. The word *latrine* has very negative associations (flies, odors, inconvenience); people view latrines as symbols of poverty and social exclusion. They do not perceive that the benefits of installing them are worth the costs. In contrast, they view bathrooms as clean and hygienic, easy to clean, and comfortable to use and associate them with modernity and progress; having a toilet conveys a sense of status.

What Poor Households Would Like

Poor households are looking for a much broader and better sanitation experience, one with options for good-quality products, offered by an accessible and credible person as part of a larger service package (including maintenance). Table 14.3 summarizes the features they are looking for.

The desire for good-quality products that are easy to maintain, accessibility of service, credibility and choice, and completeness of service are borne out by the experience of poor households that were satisfied with their sanitation solutions. In Bangladesh, where satisfaction rates among focus group participants were higher than in the other countries, the most important factors influencing the decision to buy latrines from the local entrepreneur were price, reputation, assurance of after-sales services, easy transportation, and the variety of latrine types.

Table 14.3 What Poor People in Bangladesh, Indonesia, Peru, and Tanzania Look for in a Sanitation Solution

Country	Ideal product qualities	Buying experience	Add-ons desired
Bangladesh	Ease of transport, quality plastic pan, raised platform and superstructure with bamboo walls and corrugated iron sheet roofs	Would buy from local prefabricated concrete component manufacturers that provide them options	Transport of slab, emptying of pit, repair and maintenance, warranty
Indonesia	Durable facility, ease of maintenance	Would like options to be presented	Warranty of installation, septic tank emptying service
Peru	Do not want a dry system, even if water supply is not available; want permanent, not temporary, solution and ability to purchase materials and build incrementally with a clear vision of final product	Would like guidance on price	Not specified
Tanzania	Durable and long-lasting; pit should not fill up easily; door and lid for hole	Would like products to be available nearby and sold by technician who is honest about pricing	All-in provision of labor, materials, emptying of pit or moving.

Enterprises seem to be aware of how important these aspects are to their customers, as data from Peru reveal (figure 14.3).

Lack of information or misinformation among focus group participants was apparent. In Indonesia, for example, people wondered how often a septic tank needed to be emptied and whether it was more cost-effective to dig a new pit or empty an existing one. People in Indonesia and Peru lacked knowledge about how to build their own facilities, where to go for help, and what standards products were expected to meet. Lack of adequate information on design options, costs, capacity, and builders may help explain households' unwillingness to pay for products and services. In Bangladesh, some enterprises with more skilled and entrepreneurial proprietors have been able to address some of these issues.

Many poor households are unable to build systems because of lack of land or water supply. Among people who can have facilities, motivations for investing include the ease of achieving a satisfactory solution; aspirational drivers such as modernity, comfort, dignity, and peer approval; and awareness of why sanitation is important (table 14.4).

Women's Role in Decision Making about Sanitation

In all four countries, women placed a higher priority on sanitation than did men, partly out of concern for their children. Focus group discussions revealed the role of women in initiating decisions about sanitation (figure 14.4). In Tanzania, women usually initiate the discussion about building a latrine or toilet.

**Figure 14.3	Factors Enterprises in Peru Think Consumers Consider Important in Purchasing Improved Sanitation**

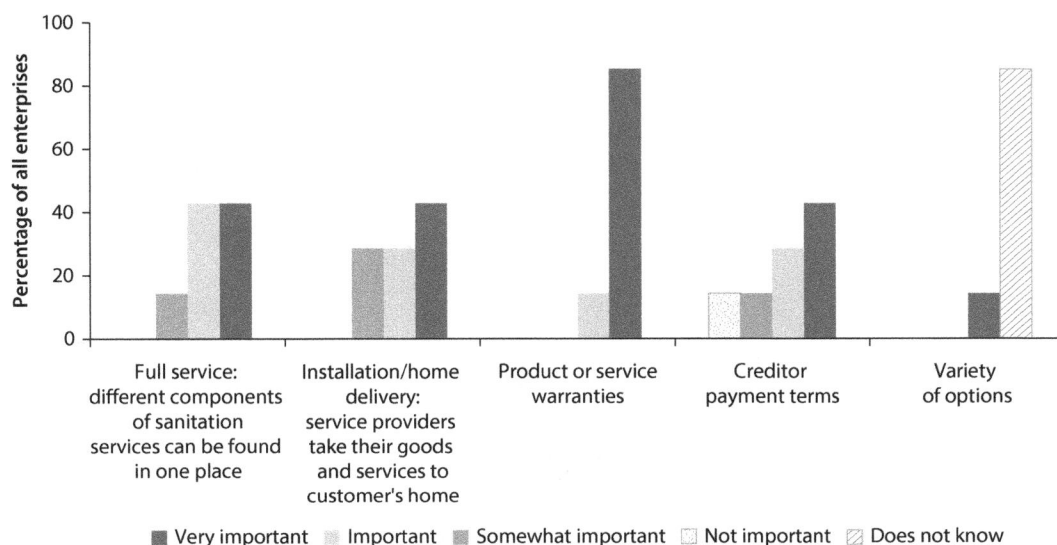

■ Very important	▓ Important	▨ Somewhat important	░ Not important	▧ Does not know

Source: IMASEN and Ausejo Consulting 2013.

**Table 14.4	Nonprice Factors Motivating Households in Bangladesh, Indonesia, Peru, and Tanzania to Purchase Improved Sanitation, 2012**

Country	Factor
Bangladesh	Women's privacy, health, saving of treatment costs, dignity, improved standard of living, approval of peers
Indonesia	Availability of land, durability (and length of time before pits or tanks need emptying), avoidance of contamination of water sources, comfort, health, ease of maintenance
Peru	Comfort, modernity, hygiene
Tanzania	Safety, durability, ease of use and maintenance, appealing product design, ease of access, hygiene and health, ease of transport and installation, privacy, modernity

Men do not seem to value household latrines or toilets because they are away from home most of the day and can use facilities outside their homes (in schools or towns, for example). Both men and women agreed the women make the decision about building or not a building a latrine or toilet. The move toward improved sanitation was viewed as a joint decision, however, with the woman acting as the initiator and the man as the implementer.

In contrast, women play a minor role in the sanitation supply business. The average share of women in full-time employment in the enterprises surveyed was 6 percent in Bangladesh, 9 percent in Indonesia, and 19 percent in Tanzania; the shares of women in part-time employment were 17 percent in Bangladesh, 38 percent in Indonesia, and 14 percent in Tanzania.

Figure 14.4 Women's Role in Sanitation in Bangladesh

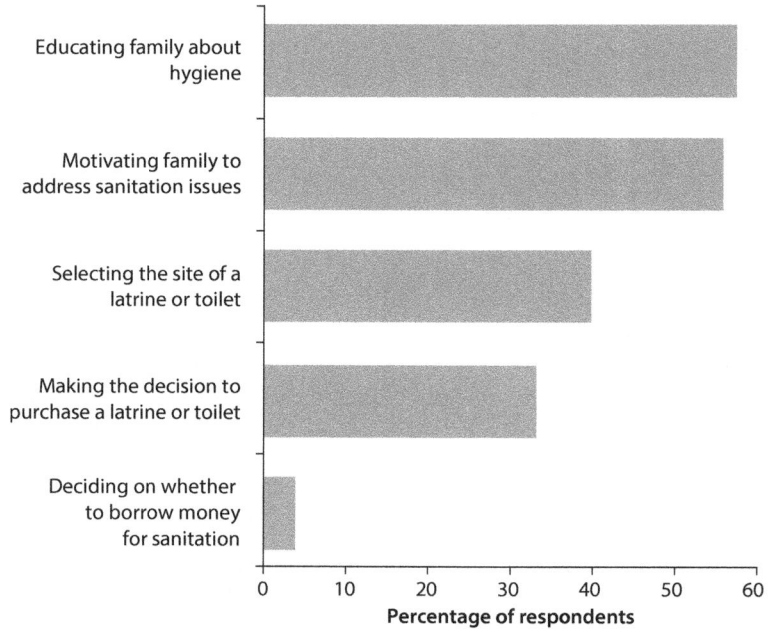

Source: DevCon 2013.

Note

1. Among the households covered by the focus group discussions in Bangladesh, nearly 80 percent of which were classified as poor or extremely poor, 82 percent had a mobile phone and 25 percent had more than one mobile phone.

References

DevCon (DevConsultants Limited). 2013. *Sanitation Bangladesh: Global Study for the Expansion of Domestic Private Sector Participation in the Water and Sanitation Market.* Dhaka: DevCon.

IMASEN and Ausejo Consulting. 2013. *Peru Country Report: Global Study for the Expansion of Domestic Private Sector Participation in the Water and Sanitation Market.* Lima: IMASEN and Ausejo Consulting.

CHAPTER 15

How Is On-Site Sanitation Supplied?

Households interested in purchasing on-site sanitation are often faced with an uncoordinated supply chain characterized by microenterprises with limited geographical reach and low turnover, selling generic items with little or no branding, quality assurance, or organized marketing. The enterprises that deal directly with households lack the capacity and resources to identify and act on opportunities to provide value-adding services to attract customers, and current technologies offer no avenues to reduce prices to stimulate demand.

Enterprise Characteristics

Enterprises selling on-site sanitation services to households are very small-scale operations. They are usually informal, have limited investment, do not keep financial records, do very little marketing, and rely on a fragmented and costly supply chain in which the major players do not view sanitation as an important part of their business.

Enterprise Size

Enterprises providing sanitation services directly to poor households in Bangladesh, Indonesia, and Tanzania are typically microenterprises (defined as having fewer than five employees) (figure 15.1). In contrast, in Peru, many enterprises are medium-size or large.

Scope and Scale of Activities

Across the four study countries, the main revenue-generating activity of enterprises in the sanitation sector falls into one of four main activities (table 15.1). Enterprises provide a variety of services (figure 15.2).

In Indonesia, 92 percent of enterprises' revenues came from sanitation. This figure was 56 percent in Bangladesh and 67 percent in Peru. In Tanzania, enterprises were either hardware stores selling a range of products or masons that took on a range of building tasks. For both, sanitation represented only a small share of their business.

Figure 15.1 Average Number of Employees of Sanitation Enterprises in Bangladesh, Indonesia, Peru, and Tanzania, 2012

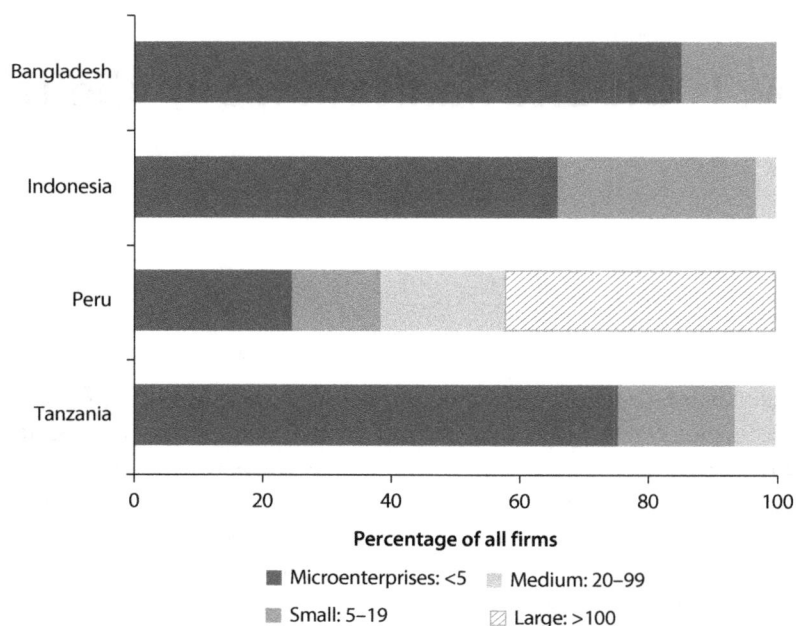

Percentage of all firms

■ Microenterprises: <5   ▨ Medium: 20–99
■ Small: 5–19   ▨ Large: >100

Source: DevCon 2013.

Table 15.1 Main Revenue-Generating Activity of Enterprises in the Sanitation Sector in Bangladesh, Indonesia, Peru, and Tanzania

Subsector	Activities
Product sales	Hardware stores that retail components and raw materials, such as cement, bricks, and reinforcing wire. Some of these stores install facilities, directly or by contracting semiskilled contractors or laborers.
Manufacture of prefabricated cement products	Enterprises that cast concrete products, such as rings, and slabs, as well as other items, such as tiles. Most are involved in building latrines and toilets; some sell products to contractors and households.
Labor and masonry	Individuals who contract directly with households or through other players, such as construction enterprises, to undertake on-site construction activities.
Pit emptying and septage removal	In Indonesia, contractors that operate trucks and pumps. Some lease equipment from and operate for local governments.

Most enterprises reached markets in a single town or district. The only exceptions were national construction companies and regional water and sewerage operators in Peru. The vast majority of enterprises operated either at the subdistrict level (all enterprises in Bangladesh; masons and most hardware stores in Tanzania) or at the level of a district or town (some hardware stores and construction enterprises and pit-emptying enterprises in Indonesia and Peru).

The scale of operations is very small (table 15.2). The average number of toilets constructed the previous year ranged between 25 (Tanzania) and

Figure 15.2 Sanitation Services Provided Directly to Households in Bangladesh, Indonesia, Peru, and Tanzania

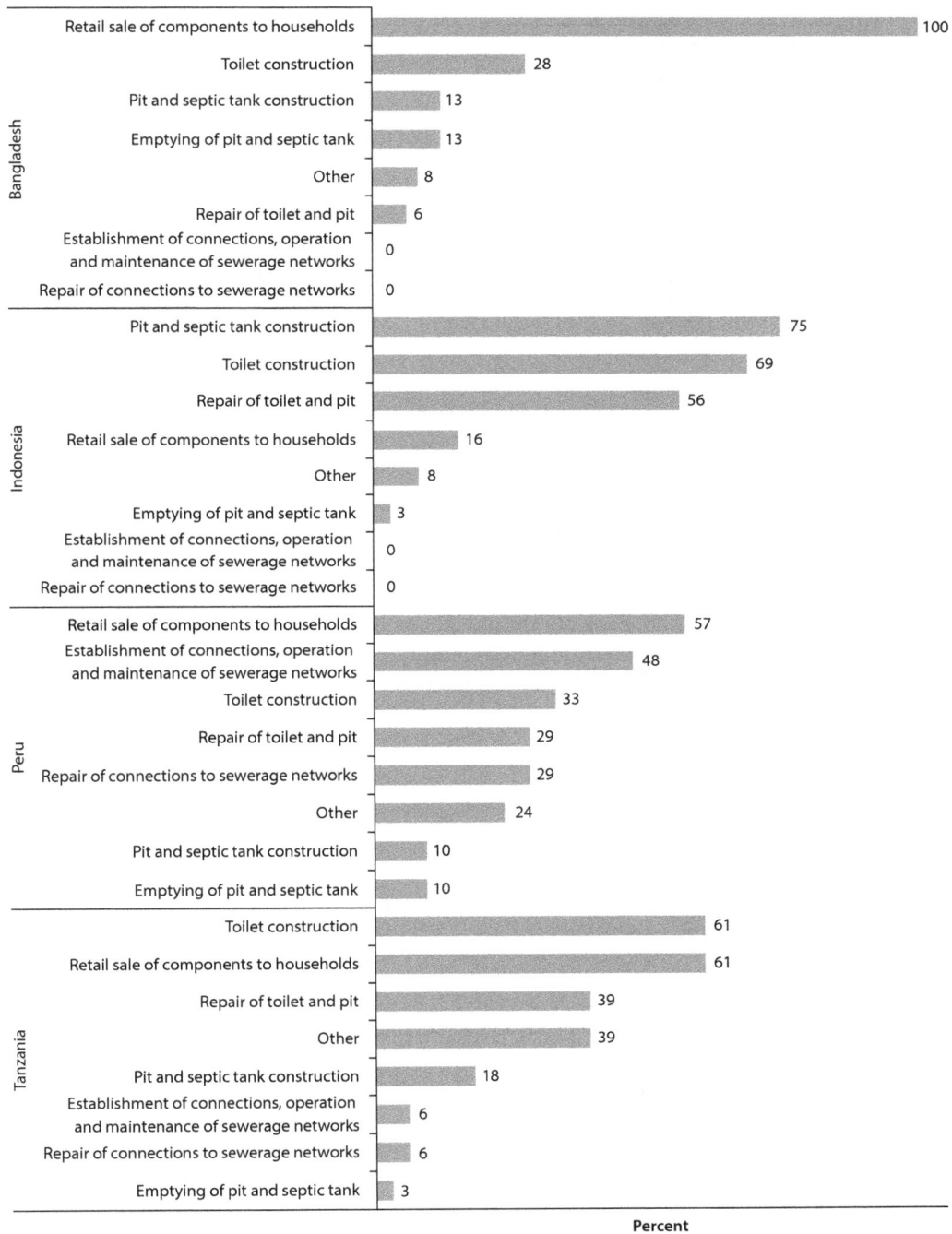

Bangladesh
- Retail sale of components to households: 100
- Toilet construction: 28
- Pit and septic tank construction: 13
- Emptying of pit and septic tank: 13
- Other: 8
- Repair of toilet and pit: 6
- Establishment of connections, operation and maintenance of sewerage networks: 0
- Repair of connections to sewerage networks: 0

Indonesia
- Pit and septic tank construction: 75
- Toilet construction: 69
- Repair of toilet and pit: 56
- Retail sale of components to households: 16
- Other: 8
- Emptying of pit and septic tank: 3
- Establishment of connections, operation and maintenance of sewerage networks: 0
- Repair of connections to sewerage networks: 0

Peru
- Retail sale of components to households: 57
- Establishment of connections, operation and maintenance of sewerage networks: 48
- Toilet construction: 33
- Repair of toilet and pit: 29
- Repair of connections to sewerage networks: 29
- Other: 24
- Pit and septic tank construction: 10
- Emptying of pit and septic tank: 10

Tanzania
- Toilet construction: 61
- Retail sale of components to households: 61
- Repair of toilet and pit: 39
- Other: 39
- Pit and septic tank construction: 18
- Establishment of connections, operation and maintenance of sewerage networks: 6
- Repair of connections to sewerage networks: 6
- Emptying of pit and septic tank: 3

Percent

Table 15.2 Average Annual Scale of Sanitation Operations in Bangladesh, Indonesia, Peru, and Tanzania, 2012

	Bangladesh	Indonesia	Peru	Tanzania
Production or services				
Number of toilets/latrines constructed	97	98	81	25
Number of pits/septic tanks constructed	138	99	—	2
Number of toilets connected to sewerage network	—	—	—	4
Number of toilets/pits/septic tanks repaired	417	26	—	22
Number of households whose pits/septic tanks were emptied	500	229	15	—
Sales				
Value of sanitation accessories sold (dollars)	5,849	1,820	22,679	—

Note: — = not available.

98 (Indonesia); the average number of households whose pits or septic tanks were emptied ranged between 0 (Tanzania) and 500 (Bangladesh).

Construction and servicing (emptying) of sanitation is currently a low-volume business. Enterprises in Bangladesh constructed 8–11 toilets or pits a month, and enterprises in Peru handled about seven installations a month. Hardware stores in Tanzania that were engaged in sanitation construction serviced 35 households, and masons installed 10 units a month. The average number of pits emptied a month was 40 in Bangladesh and 19 in Indonesia (in contrast, specialist pit-emptying enterprises in Indonesia emptied an average of 60 pits a month). In Indonesia, enterprises specializing in construction (as opposed to pit emptying) built an average of 14 latrines and pits/septic tanks a month.

Formality

The majority of surveyed enterprises involved in on-site sanitation have very simple business structures (sole proprietorships) or no formal constitution as businesses (figure 15.3). Most enterprises were registered only with local government authorities. Only in Peru are significant numbers formally registered as limited liability companies.

Capitalization

Capital intensity varies—hardware stores and pit-emptying operations have much more fixed and working capital than do masons and construction enterprises—but the overall level of investment is small (table 15.3).

Business Models

Profitability

The enterprises surveyed did not keep detailed financial records, making it difficult to review only their sanitation operations. Several findings nevertheless emerge.

Figure 15.3 Legal Form of Sanitation Enterprises in Bangladesh, Indonesia, Peru, and Tanzania, 2012

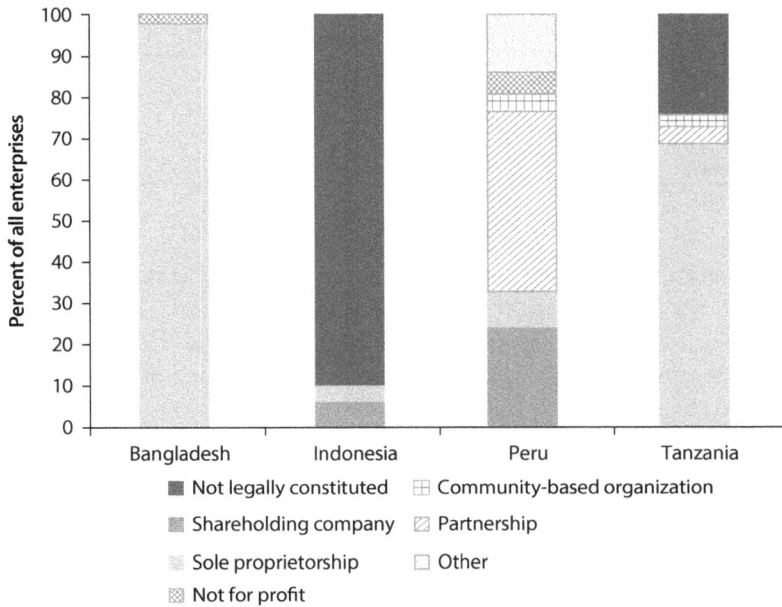

Table 15.3 Average Investment by Sanitation Enterprises in Bangladesh, Indonesia, Peru, and Tanzania, since Inception

Dollars, except where otherwise indicated

Item	Bangladesh	Indonesia	Peru	Tanzania
Number of enterprises reporting	34	32	3	13
Minimum investment	335	21	2,222	286
Maximum investment	24,450	60,963	506,556	31,807
Average investment	5,310	4,663	173,296	11,747

Note: Figures show total investment in fixed and working capital from inception to 2011. The sample in Peru included one atypical enterprise, which had invested more than $500,000.

More than 90 percent of enterprises covered their operating costs. Annual revenues ranged from $9,000 to $11,000 for enterprises in Tanzania and Bangladesh, respectively, to more than $40,000 for enterprises in Indonesia (table 15.4).

On a per unit basis, enterprises generate adequate profit margins. In Indonesia, average estimated margins were about 46 percent of sales for pit-emptying enterprises and 37 percent for construction enterprises; margins were smaller in Bangladesh (about 13 percent). Only in Tanzania were margins very low (2.6 percent). (Given the large number of family-owned operations, measured profits may well include implied returns to family labor as well as returns to capital and ownership.) Low profitability appears to reflect low volumes. Enterprises could substantially increase their margins by moving from the manufacture and

Table 15.4 Average Annual Revenues and Earnings by Sanitation Enterprises in Bangladesh, Indonesia, and Tanzania, 2011
Dollars

Country	Average annual revenues	Net profit
Bangladesh	11,000	6,300
Indonesia, construction	46,000	33,000
Indonesia, pit emptying	50,000	43,000
Tanzania	9,000	5,000

sale of sanitation components to the manufacture and installation of services (essentially adding labor), as the example in table 15.5 indicates.

In Indonesia, some enterprises have addressed pricing and ability-to-pay issues by selling modular units that can be upgraded as needs and ability to pay evolve (table 15.6).

Cost Structure

Enterprises are not likely to be able to increase margins by lowering costs, because 60–80 percent of costs are linked to materials (mainly iron, cement, and sand) whose prices cannot be negotiated. The typical enterprise buys these inputs in small quantities (in the case of masons, often only after a contract has been accepted). The materials are typically produced (or imported) by large enterprises operating capital-intensive plants located close to raw materials or ports. They are (relatively) low value to weight/volume commodities, for which transport costs can be significant. Capital constraints and rational risk aversion mean sanitation enterprises are unable or unwilling to benefit from bulk purchasing discounts or bypass intermediate players in the distribution chain. Moreover, the technologies for making the sanitation solutions are fixed-proportion technologies. There is no scope for substituting cheaper inputs or reducing input volumes without seriously reducing the integrity and durability of the product.

Embedded in the cost of input supplies and sanitation construction is transport and distance from the work site, which add an estimated 10–20 percent to the price at each step from the wholesaler, regional hardware store, and local retailer. Theft and breakage in transport also raises costs.

Some indication of the importance of transport is suggested by the data in table 15.7, which shows how far workers in Tanzania need to travel to install sanitation devices. One householder interviewed described the problem as follows: "The hardware shops are far away from here …. If they could be nearer, one could buy even a bag of cement per month and put it inside the house. Because of the distance, [it is] inconvenient to pay a fare to go and return compared to any gain."

Because the supply chain is fragmented, much effort is spent aggregating materials for construction. In Tanzania, for example, masons building latrines spend about 70 percent of their time organizing material supply.

Table 15.5 Profits of Bangladeshi Enterprise from Selling Pit Materials and Components and Installing Twin Pit Toilet

Tk, except where otherwise indicated

Item	Sale of pit materials and components	Installation of twin pit toilet
Sale price	600	4,400
Cost	510	2,635
Operation profit	90	1,765
Profit margin (percent)	18	67

Source: WSP 2012.
Note: Upgrading a product (rather than buying the upgraded product at the start) adds a day of labor.

Table 15.6 Modular Toilet Designs in Indonesia

Toilet type	Description	Cost (dollars)
WC Ekonomis	Branded ceramic closet, slab, concrete ring, cover, two days labor	60
WC Tumbuh Sehat	Branded ceramic closet, slab, one-meter pit, one day labor (does not include cover), upgradable to WC Sehat Murah Sumade	26
WC Tumbuh Sehat	Branded ceramic closet, slab, one day labor (does not include ring), upgradable to WC Ekonomis	18
WC Sehat Murah Sumade	Branded ceramic closet, slab, concrete ring, one-meter pit, cover, two days labor	85

Table 15.7 Travel Times in Tanzania to Reach Households for Latrine Construction

Minutes

Method of transport	Average travel time	Minimum travel time	Maximum travel time
Foot	51	10	360
Bicycle	110	45	300
Vehicle	208	15	1,440

Source: PATH 2012.

Marketing

Enterprises rely primarily on governments and nongovernmental organizations for information about sanitation technologies and their characteristics and in marketing the benefits of improved sanitation. They do little to market their services themselves. They focus on a limited geographical area and rely on referrals and walk-ins.

In Bangladesh, almost half of surveyed enterprises engaged in no marketing at all, and nearly 90 percent relied on word of mouth to inform customers of their products and services. Of the enterprises that did not market, nearly all said that they had enough business, so that marketing was not needed. In Indonesia, only about 20 percent of enterprises engaged in some form of marketing or advertising, and just 15 percent used sales agents. In Tanzania, only 30 percent of surveyed enterprises reported any marketing, and almost 80 percent relied entirely on word of mouth.

Table 15.8 Supply Chain Constraints in Tanzania

Actor	Characteristics	Customers	Products	Constraints
Input suppliers				
Manufacturers, importers	Located in major towns, well capitalized	Wholesalers	Construction materials	Passive sales approach
		Retailers	High degree of specialization among manufacturers	Far removed from end customer
		Large construction projects	Wider product range among importers	Focus on immediate customers for construction commodities
		Households (very small amounts)		Little knowledge of end use
				Highly specialized
Distributors				
Wholesalers, retailers	In regional towns, formally registered, sufficiently capitalized, family owned	Smaller retailers	Construction materials, including latrine components, broken up from bulk supplies to sell in smaller lots	Products not sanitation specific
		Construction projects	Tools and equipment	Passive sales approach
		Households (small amounts)		Limited market information
				Sanitation only small part of business
Local retailers				
Hardware stores, retail shops	At ward and village level, sole proprietorships, thinly capitalized	Small construction projects	Construction materials sold in very small lots	Very thinly capitalized
		Households	Household consumer goods	On-site marketing because of localized market
			Farming inputs, especially fertilizer	Small share of sales from sanitation products
				Owners often engaged in other income-generating activities, such as farming
				High cost of transport for goods

table continues next page

114

Table 15.8 Supply Chain Constraints in Tanzania (continued)

Actor	Characteristics	Customers	Products	Constraints
Masons				
Construction workers	Little or no capital invested; village-level market; face heavy demand for urban and large-scale construction projects; highly mobile	Households	Services only	Passive sales approach
		Construction projects	Slab production, construction advice, building and construction	Limited technical and business knowledge
			Latrine construction	Other income-generating activities undertaken
				High cost to mobilize materials and get to site
				Lack of capital

Source: WSP 2013.

Supply Chains

One factor preventing better alternatives from being offered to potential customers is the fragmented supply chain, in which independent enterprises manufacture or supply one or more types of materials or pieces of equipment (table 15.8). For most manufacturers, importers, and retailers, sanitation represents a very small part of their total sales. The availability of construction materials is thus driven by the demand for construction activities in other sectors. Materials and equipment that are part of separate supply chains converge at various levels of the supply chain (wholesale, retail, and consumer levels).

Households typically help construct their own latrines and toilets. But particularly where they do not have a latrine or toilet in their home, purchasing an improved sanitation solution can be challenging, because they often have to aggregate components and coordinate construction. Enterprises make little effort to market sanitation solutions or to improve coordination, exert quality control, or reduce costs within the supply chain. Actors that have the resources to address these challenges do not see sanitation as an important part of their market; the enterprises closest to the market are very small and constrained in geographic reach. Few of these enterprises specialize in sanitation services. They find it hard to signal any unique quality of service outside of the immediate vicinity where reputation is attested to by word of mouth.

Most players in the supply chain take a very passive stance toward sanitation. The technologies used do not lend themselves to economies of scale or scope in production or stock management or to any kind of branding that might make marketing useful. There are no large well-resourced players for whom on-site sanitation is a large enough market to warrant intensive efforts to market solutions or coordinate activities across the supply chain.

Enterprises catering to poor households deliver value and are generating profits, but they find it difficult to scale up, horizontally or vertically, to offer compelling products and services to the poor. Where labor is the main input driver, some horizontal integration is possible by moving from low-technology manufacturing to semiskilled installation or pit emptying. Where the input requirement is capital, many enterprises will be limited in their capabilities.

References

DevCon (DevConsultants Limited). 2013. *Sanitation Bangladesh: Global Study for the Expansion of Domestic Private Sector Participation in the Water and Sanitation Market.* Dhaka: DevCon.

PATH. 2012. *Market Assessment of Domestic Private Sector Provision of Household Sanitation in Tanzania, Final Country Report.* Seattle and Washington, DC: PATH.

WSP (Water and Sanitation Program). 2012. *What Does It Take to Scale Up Rural Sanitation?* Washington, DC: World Bank. http://www.wsp.org/wsp/sites/wsp.org/files/publications/WSP-What-does-it-take-to-scale-up-rural-sanitation.pdf.

———. 2013. "Sanitation Supply Chain Analysis." Internal working paper. Nairobi: World Bank.

CHAPTER 16

Are Enterprises Interested in Increasing Investment and Serving the Poor?

Expanding coverage of improved sanitation among poor households will involve increasing production capacity, moving capacity to areas where demand exists, investing in marketing, bundling products and services, and developing and adopting new materials and technologies. This expansion may come through new players or additional investment by current suppliers. Whatever the source, it will require investment.

Intentions to Invest

Intentions to invest differ across countries. In Indonesia, 75 percent of enterprises said they planned to invest in the coming three years. In contrast, just 33 percent of enterprises in Bangladesh and less than half of enterprises in Tanzania (48 percent of hardware store owners and 38 percent of masons) intended to do so. In Peru, where most enterprises do not regard sanitation as a primary business, 87 percent were intending to expand their sanitation-related activities over the next three years. In Bangladesh, most enterprises considering investment wanted to invest in stocking and expanding sales (80 percent) and manufacturing (77 percent) of latrine and toilet components. Few had interest in expanding into installation or repair of latrines and toilets or other sanitation-related businesses.

In Indonesia, 85 percent of enterprises were planning to increase the range of sanitation-related services, responding to signals from customers about the desire for service bundling. Enterprises involved primarily in construction of latrines and toilets were considering getting involved in designing and consulting on sanitation systems (70 percent), selling sanitation-related consumer products (45 percent), and treating and disposing of wastewater (40 percent). The optimism of construction enterprises is reflected in the number of households they expected to serve the following year. Half of respondents in the construction business were sure that they would serve more than 500 households the following year—a quadrupling of volume from the average of 143 households at the time of the interviews. Pit-emptying respondents expected only a modest increase in the number of customers.

In Tanzania, nearly all enterprises contemplating investment were thinking of expanding their current lines of business rather than moving into other sanitation-related activities.

Perceived Risks

Asked to name obstacles to investment, 50 percent of enterprises in Indonesia and 75 percent in Tanzania indicated that the level of demand was a major concern (figure 16.1). In Bangladesh, a significant proportion of enterprises worried about finding reliable workers to manage additional business.

Figure 16.1 Enterprises' Assessment of Obstacles to Investment in Sanitation in Bangladesh, Indonesia, and Tanzania

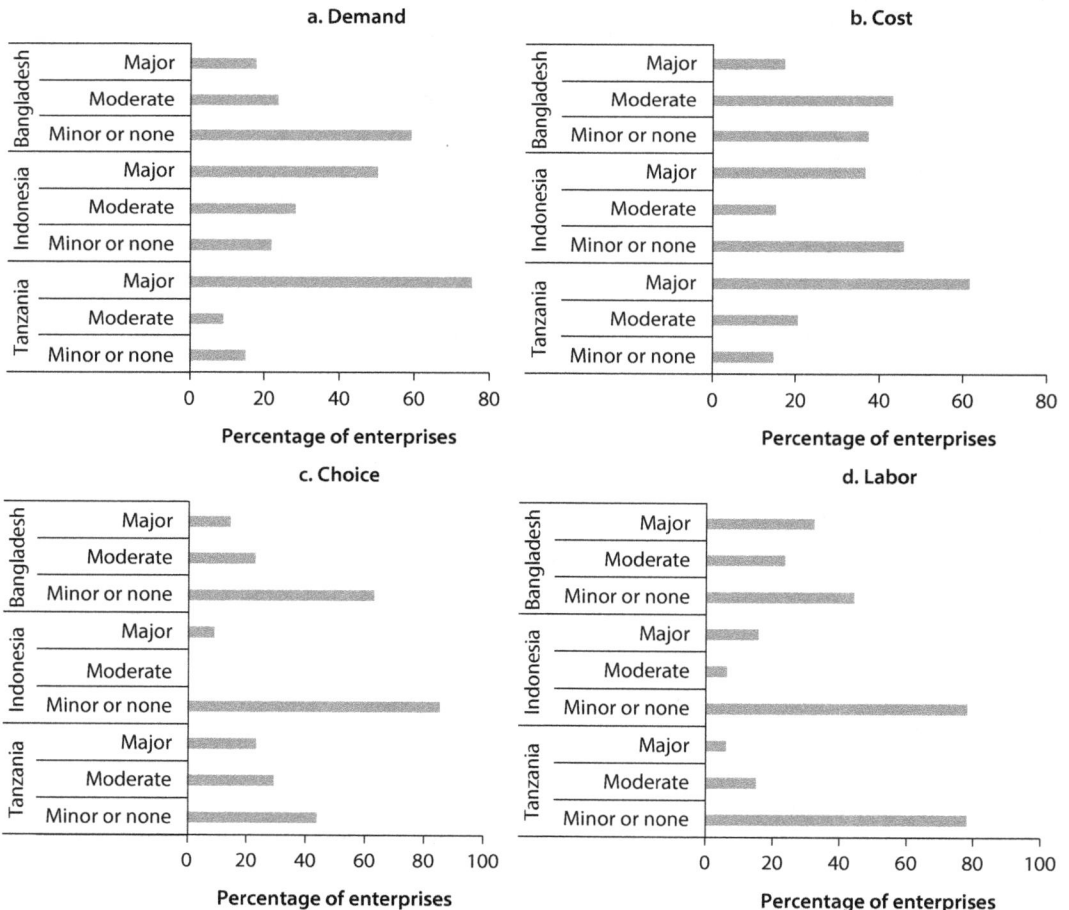

In Tanzania, 63 percent of enterprises were concerned that investment would be too costly to be profitable. In Tanzania and Bangladesh, around 50 percent of enterprises considered that difficulty in choosing what to invest in was a moderate or major problem.

Perceptions of the Poor as a Target Market

Perceptions of the poor as an attractive customer segment vary. In Bangladesh and Indonesia, more than 60 percent of enterprises agreed or strongly agreed that the poor were target customers for them. In contrast, just 48 percent of respondents in Tanzania did so, with a third strongly disagreeing that this was the case (figure 16.2). More than three-quarters of Bangladeshi enterprises indicated that the poor do not pay on time, a view shared by smaller majorities in Indonesia (54 percent) and Tanzania (63 percent). In Tanzania, respondents recognized that a significant number of households lacked improved sanitation. Masons were much more likely than hardware stores to target the poor. Less than a quarter of hardware store owners or managers agreed or strongly agreed

Figure 16.2 Enterprises' Perceptions of the Poor as Target Customers and Assessment of Their Attitudes toward Sanitation in Bangladesh, Indonesia, and Tanzania

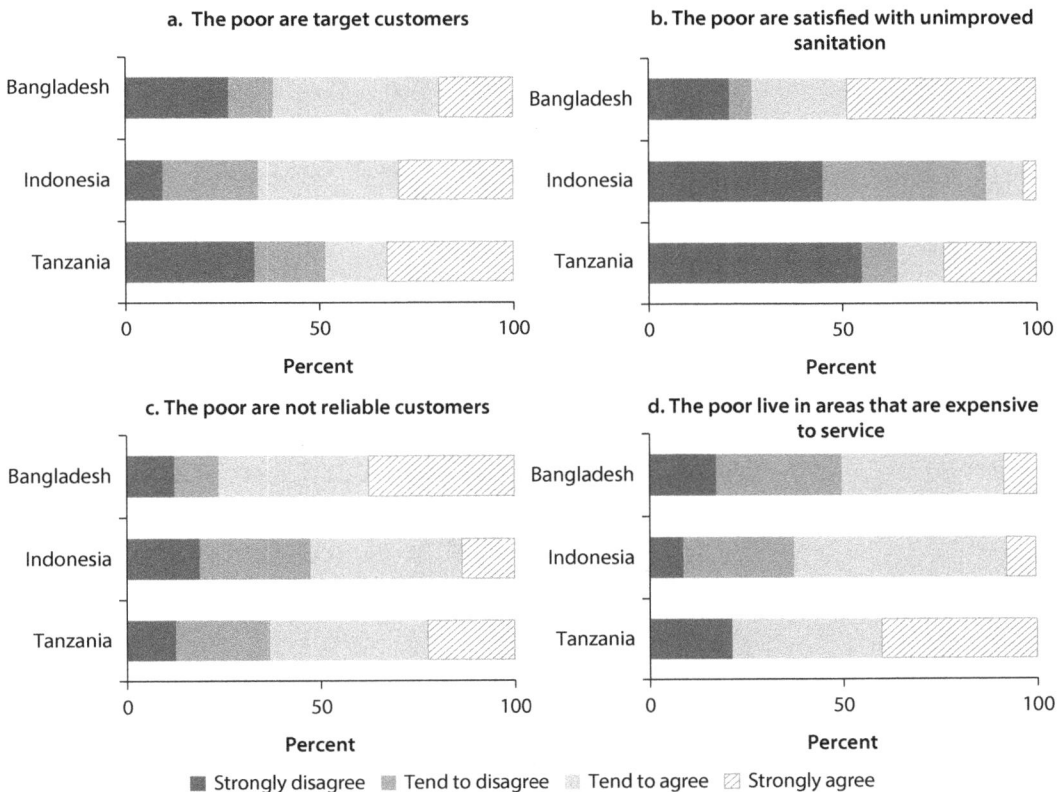

that poor households are their major target customers. In contrast, 91 percent of masons agreed or strongly agreed that the poor were a major target group.

A majority of enterprises in all three countries (77 percent in Bangladesh, 54 percent in Indonesia, and 63 percent in Tanzania) believe that the poor are not reliable customers in terms of paying on time. A majority of Peruvian enterprises (57 percent) disagreed that a 10–20 percent price reduction would increase their sales to poor households. They agreed that the inability of poor households to make large cash outlays is the most important constraint on their ability to pay.

In Indonesia, enterprises involved in pit emptying did not view poor households as an important part of their market, because poor households tend not to have pits or septic tanks that need emptying. Construction businesses were more engaged with these households: nearly 70 percent agreed or strongly agreed that poor households were their major target customers. The majority of both pit-emptying enterprises (50 percent) and construction companies (70 percent) agreed that the poor lived in areas that would be harder, and therefore more costly, to reach. About half of these enterprises thought that poor customers would not make timely payments.

More than three-quarters of enterprises in Tanzania indicated that the poor lived in areas that were expensive to service because of transport and infrastructure problems.

CHAPTER 17

Is the Investment Climate Limiting Private Sector Involvement?

Government policy and practice, the quality of infrastructure, and access to finance shape the way enterprises perceive the trade-off between risk and return when considering expanding their business. All of these aspects of the investment climate affect the sanitation sector.

Government Policy and Practice

Table 17.1 summarizes enterprises' views on the extent to which different aspects of governance act as obstacles to doing business. It shows that enterprises in Peru and Tanzania more frequently identified governance issues as major to very severe obstacles.

Lack of Market Intelligence and Inability to Conduct Research and Development

Enterprises identified few restrictive actions arising from government policy or action that would prevent them from entering the market. The problem is therefore not what governments are doing but what they are not doing.

None of the countries had specific mechanisms or incentives set up to promote private sector entry into the market. Few enterprises in Indonesia could point to specific government programs that prioritized sanitation service delivery to the poor. In Tanzania, few enterprises could identify institutions that could address the needs of the poor in sanitation.

Two areas emerge as requiring proactive action from government if a market-based approach to sanitation service delivery is to result in widespread access by the poor: provision of market intelligence and the facilitation of entry by enterprises that have research and development (R&D) capabilities.

Even among existing enterprises, there is concern about the profitability of their planned investments and the regularity of demand by the poor. Sanitation to

Table 17.1 Enterprises' Perceptions of Governance-Related Obstacles in Bangladesh, Indonesia, Peru, and Tanzania

Percentage of enterprises identifying issue as an obstacle

Country/severity	Corruption	Unpredictability[a]	Political instability	Restrictions on entry into markets in other locations	Project development procedures
Bangladesh					
None/no view	35	32	15	53	65
Minor–moderate	50	59	56	38	35
Major–very severe	15	9	29	9	0
Indonesia					
None/no view	84	88	81	66	75
Minor–moderate	3	9	13	22	16
Major–very severe	13	3	6	13	9
Peru					
None/no view	48	33	19	52	38
Minor–moderate	19	29	43	24	33
Major–very severe	33	38	38	24	29
Tanzania					
None/no view	39	36	39	27	27
Minor–moderate	15	30	33	55	33
Major–very severe	45	33	27	18	39

a. Including lack of consistency of local government administration.

the poor is a nascent market, in which entry costs for first movers are high. Enterprises lack the analytical tools to determine the existence of a potential market that needs servicing. Without a clear idea of the volume and nature of market demand, they have no way of knowing how supply structures and offers might need to change to meet it. Lack of information probably also increases enterprises' perceptions of risk.

Between 40 and 50 percent of enterprises in Bangladesh, Indonesia, and Tanzania believe that technological improvements are necessary to better meet the needs of the poor, who often live in flood-prone or steeply sloped areas, where standard approaches do not work well (figure 17.1). Few enterprises indicated that the availability of appropriate, affordable technology would motivate them to specifically cater to poor households, however (although in Tanzania, 75 percent of masons, who engage much more directly with households than do hardware stores, agreed with the statement).

Enterprises look to government for innovation. Government should not necessarily be developing technology, but it can actively promote R&D on sanitation solutions that are suited to the living conditions and life aspirations of the poor, through grants, patent protection, contracts, and accreditation systems.

Figure 17.1 Enterprises' Views on Whether Technological Improvements Are Needed to Address Problems Where Poor Households Live

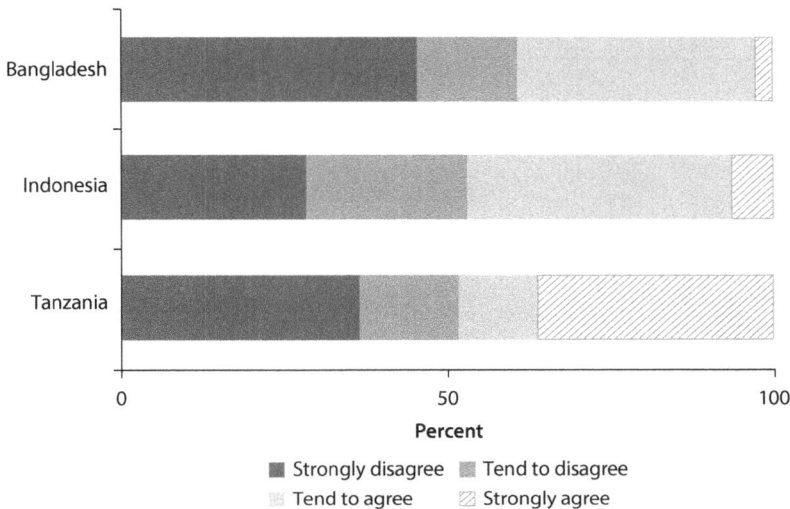

■ Strongly disagree   ■ Tend to disagree
▒ Tend to agree   ▨ Strongly agree

Bureaucracy, Uncertainty, and the "Hassle Factor"

Enterprise perspectives on the impact of corruption vary considerably. In Indonesia, where enterprises reported paying up to 9 percent of annual sales in payments to "get things done," corruption was generally not seen to be a significant problem but instead regarded as part of the rules of the game. In Tanzania, 45 percent of enterprises reported that corruption was a major to very severe obstacle, but two-thirds said they did not know how much they paid annually in informal gifts.

In Bangladesh and Tanzania, slightly more than half of enterprises had obtained some kind of business permit or license. In contrast, in Indonesia, 85 percent had not. Enterprises that had obtained a permit did so for pit emptying, which is a regulated activity. Among respondents, 80 percent in Bangladesh and 90 percent in Indonesia said that permits and the need to obtain them represented no or only a minor obstacle to doing business. In Tanzania, 24 percent of enterprises saw it as a severe obstacle.

Infrastructure

Table 17.2 summarizes enterprises' assessment of the extent to which inadequate infrastructure acts as an obstacle to their operations. In Bangladesh, more than 60 percent of enterprises viewed the water supply as a major to very severe obstacle. In all countries but Peru, a majority of enterprises viewed transport as a problem. Enterprises in Tanzania were more likely than enterprises in other countries to cite all aspects of infrastructure provision (electricity, telecommunications, water, and transport) as major to very severe obstacles to their operations.

Table 17.2 Enterprises' Perceptions of Infrastructure-Related Obstacles in Bangladesh, Indonesia, Peru, and Tanzania

Percentage of enterprises identifying issue as an obstacle

Country/severity	Electricity	Telecommunications	Water	Transport
Bangladesh				
None/no view	44	35	6	21
Minor–moderate	50	53	32	68
Major–very severe	6	12	62	12
Indonesia				
None/no view	91	53	91	50
Minor–moderate	6	31	6	16
Major–very severe	3	16	3	34
Peru				
None/no view	67	76	86	57
Minor–moderate	19	14	5	29
Major–very severe	14	10	10	14
Tanzania				
None/no view	30	39	39	21
Minor–moderate	33	21	24	36
Major–very severe	36	39	36	42

Figure 17.2 Enterprises' Assessment of Inadequate Access to Finance as an Obstacle to Current Operations in Bangladesh, Indonesia, Peru, and Tanzania

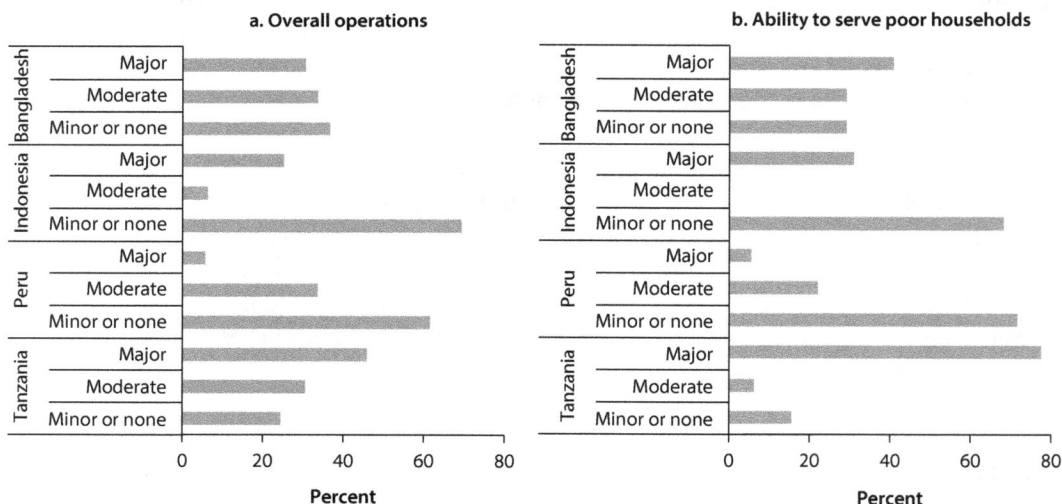

Access to Finance and Financial Services

Enterprises considered inadequate access to finance an obstacle to both their operations in general and their ability to reach poor households. A majority of enterprises in all countries except Indonesia and Peru cited access to finance as a moderate or major obstacle (figure 17.2).

Table 17.3 Percentage of Sanitation Enterprises in Bangladesh, Indonesia, Peru, and Tanzania with Bank Accounts or Line of Credit

Financial instrument	Bangladesh	Indonesia	Peru[a]	Tanzania
Bank account	47	22	67	39
Loan or line of credit from a financial institution	52	41	56	21

a. In Peru, interviewed personnel in some enterprises reported they did not know if the enterprise had a bank account (11 percent) or loan/line of credit (6 percent).

A large share of the money that enterprises invested in their sanitation business came from their own or family funds (82 percent in Bangladesh, 72 percent in Indonesia). Smaller portions came from microcredit institutions and grants from governments or nongovernmental organizations. Only a small share of financing came from commercial financial institutions (0.8 percent in Bangladesh, 11 percent in Indonesia). In Tanzania, it was harder to piece together a consistent picture of investment funding, but it appeared that most hardware stores and masons financed investment from their own resources.

Interaction with the banking system varied across the countries. The proportion of interviewed enterprises with a bank account ranged from 22 percent (Indonesia) to 47 percent (Bangladesh) (table 17.3); the proportion with a loan or line of credit ranged from 21 percent (Tanzania) to 52 percent (Bangladesh). Just 28 percent of enterprises in Bangladesh and 53 percent in Indonesia have loans or lines of credit from the banking system. Only in Peru did a majority of enterprises have bank accounts.

CHAPTER 18

Conclusions and Recommendations

Governments in Bangladesh, Indonesia, Peru, and Tanzania recognize that millions of their people in rural and semi-urban areas lack improved sanitation, but they have neither the resources nor the capacity to redress the problem directly. Together with development partners, they are looking to the domestic private sector to play a larger role in expanding access to improved sanitation.

The private sector is already serving poor households in the region. But the poor are not very interested in buying the improved on-site sanitation solutions that are being offered to them. To help the private sector improve its ability to meet the needs of this segment of the population, governments can take a variety of actions, described here.

Conclusions

The sanitation market in the four countries studied is large. Significant commercial and technological constraints prevent the domestic private sector from tapping it, however.

Market Potential Is Great

The market for improved on-site sanitation services in the four study countries is already large: supplying new systems and replacing old ones is conservatively estimated to be worth $300 million a year. The potential market is much larger: providing improved sanitation facilities to the estimated 228 million people in these countries who lack access would involve sales of at least $2.6 billion. Poor people alone would account for sales worth about $700 million, and new customers would increase the value of the replacement market to about $550 million a year. There is also significant market potential in repairing facilities and collecting and disposing of septage (in Indonesia alone the potential market for truck-based collection services is about $100 million a year).

Enterprises Are Not Offering Products and Services Households Want to Buy

The main constraint to scaling up private provision and realizing the market's potential is that businesses are not offering households products and services they want to buy. Many poor (and not-so-poor) people are unwilling to pay for the kinds of improved sanitation solutions currently available in the market. As currently structured, the supply chain delivering these solutions appears unable to offer better value.

Demand. Sanitation is a low priority for many poor households. Inability to pay does not seem to be the main reason for low demand: poor households lay out significant sums for other consumer durables, such as mobile phones. Instead, it appears that many households that are unable to afford the type of sanitation they want prefer to "make do" with inferior solutions rather than purchase what they can afford. The fact that even better-off households often lack improved sanitation means that there is not much of an "emulation" push for poorer households to move up the sanitation ladder.

Poor rural households have seasonal and unsteady cash flow and limited access to financial services that could help them smooth consumption. Weather and its impact on transport compound the seasonality of the market. Enterprises serving this market must contend with these challenges.

The problem of low prioritization and limited ability to pay is complicated because the market is heterogeneous. The drivers of household decisions to stop open defecation are likely to be different from the drivers of household decisions to move up the sanitation ladder. The strategies used to motivate households still engaged in open defecation to purchase improved sanitation may therefore have to differ from the strategies used to motivate households that already have basic sanitation. And some households not using improved sanitation may be very costly to reach because of their isolation or because cheaper technologies do not fit their circumstances.

Enterprise viability and business models. What consumers want differs from what enterprises are providing. Poor people want good-quality products that are easy to maintain, accessibility of service, credibility and choice, and completeness of service. Most private enterprises manufacture and sell components, build sanitation units, or provide pit-emptying services. Few offer a full-service option, most offer very rudimentary technologies, and the burden of coordinating construction usually falls on the consumer.

Most enterprises in the sector are profitable; enterprises in Indonesia and Peru in particular have the potential to increase their profits through value-adding. But the industry is characterized by very localized microenterprises with low turnover. The prevailing technology is generic and focused on manufacture by microenterprises; it does not lend itself to branding or coordinated marketing, and there are few opportunities to reduce costs. Few enterprises invest in marketing to increase their sales. Even fewer have the business skills to figure out how to use labor to create more value.

e

Bundling of services may be one way in which sanitation enterprises could exploit their knowledge of the market. Many enterprises recognize that bundling and expanding the scope of their activities is important to their customers. But like other ways of expanding business, adopting bundling strategies or pursuing more nuanced marketing activities involves investment, which enterprises do not appear interested in making.

One reason that the industry is not supplying products people want to buy is that the development of sanitation solutions has traditionally been seen as the preserve of the public sector, aided by nongovernmental organizations. The absence of efforts to develop and market alternative solutions also reflects the existing industry structure.

Most players in the supply chain take a very passive stance toward sanitation. There are no large, well-resourced players for whom on-site sanitation is a big enough market to warrant intensive efforts to market solutions or to coordinate activities across the supply chain. For their part, enterprises closest to the market are very small and constrained in their geographical reach. Few agents specialize in the provision of sanitation services. Most enterprises in the sector are either hardware stores or concrete fabricators, for whom sanitation makes up a small share of their business.

Attitudes toward investment and serving poor customers. Expanding coverage of improved sanitation among poor households will require expanding production capacity, relocating capacity to areas where demand exists, investing in marketing, bundling products and services, and developing and adopting new materials and technologies.

Enterprises recognize that the market for improved on-site sanitation will continue to grow, but they are concerned about the regularity of demand. A significant number of Indonesian enterprises were planning to expand the range of sanitation-related services they offered, responding to signals from customers about their desire for service bundling. In contrast, in Bangladesh, enterprises contemplating investment focused on expanding the scale of what they currently do: manufacturing and selling latrine and toilet components. Few had any interest in expanding into installation and repair of latrines and toilets or other sanitation-related business lines. The same attitude was evident in enterprises in Tanzania.

Perceptions of the poor as an attractive customer segment vary. In Bangladesh and Indonesia, more than 60 percent of enterprises agreed or strongly agreed that the poor were target customers for them. This figure was just 48 percent in Tanzania, where a third of all respondents strongly disagreed that this was the case. More than three-quarters of Bangladeshi enterprises indicated that the poor do not pay on time, a view shared by smaller majorities in Indonesia (54 percent) and Tanzania (63 percent).

In Tanzania, more than three-quarters of enterprises believe that the poor live in areas that are expensive to service because of transport and infrastructure problems.

A Weak Investment Climate Is Constraining Investment

Despite a variety of high-level strategies, plans, and statements of intent, central and lower-level governments seem to have little impact on private provision of sanitation: enterprises are typically unaware of national policy, and implementation by local-level governments is undirected and poorly funded.

Where governments have been involved in the direct supply of sanitation services to poor households, the top-down approach has not been very successful. But government provision and subsidies do not seem to be a significant source of distortion to the market. Most enterprises that provide sanitation services to households are typically too small and localized to be affected by constraints that affect the formal business sector.

Enterprises in the sector believe that governments should concentrate on addressing the market imperfections related to households' understanding of the benefits of improved sanitation and the nature of on-site solutions and on promoting the entry of enterprises able to undertake transformative research and development on new technologies and materials not within the capacity of present players. They indicate that the quality of transport infrastructure is an obstacle to increased investments. Tanzanian enterprises also cite obstacles in other infrastructure areas, and enterprises in Bangladesh identify water supply as an issue. Access to finance and financial services is low except in Peru, reducing enterprises' ability to invest and cater to the poor.

Recommendations

Scaling up private sector provision of improved sanitation to the poor requires addressing the commercial constraints that confront the sector. These constraints are inherent in the technologies used and the supply chains that support service provision.

Governments, development partners, and the business community could help relax these constraints in a variety of ways (table 18.1). They could encourage larger businesses and funders of sanitation to develop technologies with more consumer appeal; help reduce distribution costs; inject more proactive and commercial coordination into the supply chain; and help develop financial products that would enable poor households to manage the upfront costs of purchasing latrines, toilets, and septic tanks. Over the longer term, they could solve some of the infrastructural problems that raise the costs of connecting rural markets to urban centers of production of components and materials.

The study's recommendations focus primarily on the constraints to expanded private provision of on-site sanitation inherent in current technologies and in the supply chains that support service provision. It is these constraints that lead enterprises to offer products and services households are not very interested in buying.

Table 18.1 Policy Recommendations for Increasing the Provision of Improved Sanitation to the Poor

Policy goal	Recommended action	Actor
Stimulate demand by the poor		
Enhance consumer awareness	1. Improve household understanding of improved sanitation: • Complement private marketing of sanitation solutions to fill gaps in community understanding and address misinformation about the capabilities and maintenance requirements of improved on-site sanitation. • Develop education and awareness campaigns directly targeting households that already have some kind of sanitation, to complement campaigns targeting open defecation and address limited household understanding of the characteristics of improved sanitation systems. • Ensure that these campaigns address the gender dimensions of sanitation awareness and decision making where appropriate.	Governments, development partners
Improve affordability	2. Smooth and subsidize poor household sanitation expenditures: • Use instruments to help very poor households mobilize cash to pay for improved latrines/toilets that do not distort markets. • Develop and support facilities that enable payment on installment terms, intermediated either through agency arrangements with manufacturers and suppliers of components or through financial institutions that provide consumer loans to households. • Develop and finance targeted subsidies for extremely poor households or in locations where suitable technology cannot be delivered at reasonable costs.	Governments, development partners
Encourage innovation and facilitate efforts to relax business model and supply chain constraints		
Spur innovation	3. Stimulate and, if necessary, financially support the development of affordable technologies with consumer appeal: • Help develop technologies (preferably proprietary or licensable) that use materials that are light and easy to transport; easy to clean and maintain; and amenable to mass production, branding, and marketing through distribution networks coordinated and supported by manufacturers. Also help develop modular technologies that enable incremental improvements to sanitation facilities as household interest grows and as households are able to mobilize funds. • Explore options for stimulating research and development by the private sector such as through patents, contracts, and grants. • If the preferred model of commercial development and roll-out of proprietary technology is not forthcoming, consider expanding funding by the international development community of research and development to develop technologies that are appropriate for delivery through a market-based system.	Governments, development partners
Encourage larger businesses to enter the on-site sanitation sector	4. Foster the entry of well-capitalized enterprises with marketing skills to drive consumer interest and capacity to coordinate supply chains, and support installation and maintenance by small-scale local enterprises: • Support the collection and dissemination of market intelligence such as information on the size and nature of the market, including that significant segments of households above the poverty line are a part of the market. • Explore options for incentives to entry, including start-up financing and support. • Encourage the formation of associations of enterprises involved in sanitation to develop a distribution channel to the "last mile" and assist in the dissemination of market and technical information.	*Governments, development partners*

table continues next page

Table 18.1 Policy Recommendations for Increasing the Provision of Improved Sanitation to the Poor (continued)

Policy goal	Recommended action	Actor
Support quality assurance	5. Enable quality assurance and accreditation: • With the entry of larger businesses in the supply chain, assist the microenterprises at the front end to more credibly signal service quality to a larger market, and assure potential purchasers that they will get value for money and durability and continuity of service. • If capacity exists, introduce public sector certification of technologies, or government endorsement of international certification by development partners, but avoid government regulation of standards. • Facilitate industry-based accreditation systems for enterprises or solutions to enable manufacturers to offer warranties on installation.	Governments, development partners, business community
Support business capacity development	6. Help the microenterprises currently delivering the bulk of on-site solutions expand their limited business expertise so that they can better participate in an expansion of supply: • Facilitate capacity building through partnerships with larger actors in the supply chain in agency, distribution, or subcontracting networks that also address the capacity and commercial issues of the front end of the supply chain. • Develop elements of public sector sanitation marketing and education campaigns that can be used as information and marketing material by small-scale private sanitation service providers.	Governments, development partners
Improve investment climate and sectoral policy		
Facilitate private provision	7. Clearly spell out an active (rather than default) role for the private sector in government strategies and policies, and improve sector investment planning to identify markets with potential for private participation: • Detail and publicize policies to facilitate the private sector role. • Indicate responsibilities across different levels of government for implementation, especially where local governments have in-principle responsibility, mandates, and resourcing for sanitation.	Governments, development partners
Regulate septage disposal	8. Formulate practical standards and protocols for disposal of fecal sludge, and build the capacity to implement them: • Develop safe arrangements for disposal to accompany the growth of private sector pit and septic tank emptying. • Develop sites for treatment of fecal sludge, along with protocols for treatment. • Explore options for financing disposal sites, including public-private partnerships.	Governments, development partners

Appendix A

Table A.1 Demographic, Geographic, and Socioeconomic Indicators for Bangladesh, Benin, and Cambodia, 2010

Item	Bangladesh	Benin	Cambodia
Demographic and area information			
Population (millions)	150	9	15
Rural population (millions)	107	5	11
Land area (square kilometers)	14,757	110,620	176,520
Key socioeconomic indicators			
Per capita gross national income (GNI) (purchasing power parity) (current dollars)	1,649	1,580	2,080
Gini coefficient	32		44.37
Percentage of population living on less than $2 a day (purchasing power parity)	77	—	17
Percentage of population living below national poverty line	32	39 (in 2009)	—
Characteristics of natural water resource			
Annual rainfall (millimeters)	1,400	700–1,300	1,000–1,500
Rainy season	June–September	April–July and September–November	May–October
Annual renewable water resources (billion cubic meters)	105 (in 2011)	26.4	476.1 (in 1999)
Annual freshwater withdrawal (billion cubic meters)	36 (2011)	0.264	
Main water source for human consumption	Groundwater	Groundwater	—
Main river systems	Ganges-Padma, Brahmaputra-Jamuna, Meghna, Teesta	Niger	Mekong, Tonle Sap
Main problem	Arsenic, iron, salinity, overabstraction	No significant problem	Arsenic, high sediment load, turbidity, bacteriological contaminants
Improved drinking water coverage (percent)			
Whole country	81	75	64
Rural areas	80	68	58

Sources: The Gini coefficient for Cambodia and the percentage of population living below the poverty line in Benin are from country reports. All other demographic and socioeconomic data are from World Development Indicators (database) 2013. Data on annual rainfall, the rainy season, the main water source for human consumption, the main river systems, and the main problem are from country reports. Data on improved drinking water coverage are from WHO/UNICEF 2012.

Note: Data are for 2010 except where indicated otherwise. — = not available.

Table A.2 Characteristics of Surveyed Piped Water Networks and Operators in Bangladesh, Benin, and Cambodia

	Bangladesh	Benin	Cambodia
Average employment			
Number of full-time employees	2	5	3
Years in operation	5	7	8
Experience of manager (years)	5	8	11
Average production			
Length of pipe system (kilometers)	7	14	7
Total number of people served	1,504	8,023	3,177
Number of poor people served	274	4,050	—
Number of villages served	3	4	6
Number of private connections	196	23	648
Number of shared connections	18	14	0
Annual water production (cubic meters)	65,887	13,388	46,281
Annual water sales (cubic meters)	62,376	11,506	40,026
Annual water loss (cubic meters)	3,511	1,882	6,254
Production efficiency (percent)	18	30	46
Water supply availability (hours/week)	33	86	132
Sources of water for production			
Groundwater (percent)	91	100	19
Ponds/rivers, surface water (percent)	9	0	81
Billing method (percent)			
Flat rate per month	100	0	3
Based on volume consumed	0	100	97

Note: — = not available.

Table A.3 Legal Status of Water System Operators in Bangladesh, Benin, and Cambodia
Percentage of all enterprises

Item	Bangladesh	Benin	Cambodia
Registered/licensed	52	96	48
Type of firm			
Shareholding company	0	35	8
Sole proprietorship	13	58	88
Users association	56	0	0
Nonprofit organization	31	4	0
Not legally constituted	0	4	4
Part of larger firm	47	4	6
Independent firm	53	96	94
Network management			
Single-network firm	94	44	83
Multinetwork firm	6	56	17

Table A.4 Summary Characteristics of Water System Operators in Bangladesh, Benin, and Cambodia

Country	General characteristics	Business model	Finance and profitability	Outlook on and view of the poor
Bangladesh	Estimated 75 schemes operated by 32 firms. 80% are microenterprises. 85% are NGOs or users association. Few are legally constituted.	Many are NGOs. For 94%, water accounts for more than half of total revenue. 86% of operating cost is labor cost and energy cost.	$7,000–$100,000 capitalization. 75% of firms recover operating cost; 45% have negative margins after interest, tax, and amortization. 37% of firms keep financial records. Most firms have loans from commercial banks, nonbank financial institutions, or state banks.	55% plan to invest in current site enhancement. 70% do not view the poor as a target or believe the poor have equal access to water services.
Benin	Estimated 120 schemes. 60% are microenterprises; 40% are small. Firms are formally registered single proprietors or shareholding companies.	Consulting and works in water supply; operations of systems. For 44%, water accounts for more than half of total revenue. 60% of networks are operated by single-network firm.	Capitalization for office equipment. Most firms cover costs, but large percentage of gross margin is paid as government fees; revenues are typically not large enough to cover depreciation. 90% of firms keep financial records. 76% have bank accounts; 12% have loans through banks.	33% plan to invest, but only in maintenance and moveable assets; given lease contract. 36% view the poor as a target market; 60% believe that the poor do not have equal access to water services.
Cambodia	Estimated 140 licensed firms nationally. 70% are microenterprises; 30% are small. 75% not legally constituted but have license or authorization to operate.	Build-own-operate piped water supply. Modular expansion; emphasis on revenue areas and low capital investment inputs. High levels of performance, satisfaction of customers. 70% focus on water as a business. 65% of cost is energy.	$30,000–$120,000 capitalization. Most firms recover full costs, with average net margins of 23%. No firms keep financial records. 17% of firms have bank accounts; 23% have loans backed by real estate mortgages.	77% plan to invest in current site; 53% plan to invest in additional system. 93% view the poor as a target customer.

Note: NGO = nongovernmental organization.

Table A.5 Characteristics of Supply Chains in the Water Sector in Bangladesh, Benin, and Cambodia

Country	Design and technical expertise	Materials supply	Construction	Operations	Technical and business support
Bangladesh	Competitive market: about 10 large national consulting companies specialize in water supply; many consultants are also available.	Highly competitive market; local hardware suppliers carry pumps and pipes.	About 100 national construction companies specialize in construction of boreholes. Pipes are available in the capital and in all districts. Business is procured through competitive bidding under public procurement.	Operated largely by communities and NGOs, more recently by the private sector.	Limited. Provided through Department of Public Health Engineering (DPHE) or large NGOs constructing water supply systems.
Benin	Semicompetitive: 30 firms in small market; tend to work in consortium.	Most pumps and generators are imported from Europe. Pipes are imported from the region and sourced locally.	80 national and foreign firms operate independently, involved in general construction. Business is concentrated in public procurement.	Limited skilled labor. Agency risks of standpipe operators.	None
Cambodia	Local companies not mature, dominated by international firms catering to government bids; unaffordable to local private sector.	Well-developed competitive markets: three importers of pipes from China and Thailand and new local factory for high density polyethylene seem to behave competitively. Pump suppliers at local level are highly competitive.	Competitive in other markets (for example, housing), but water firms tend not to use external construction firms. Many local technicians unspecialized in water.	Excludes foreign firms.	Not developed for water sector, although competitive in more general sectors, such as accounting, information technology.

References

Country Studies

DevCon (DevConsultants Limited). 2013. *Water Supply Bangladesh: Global Study for the Expansion of Domestic Private Sector Participation in the Water and Sanitation Market.* Dhaka: DevCon.

GRET (Groupe de Recherché et d'Echanges Technologiques). 2013. *Final Report Cambodia: Global Study for the Expansion of Domestic Private Sector Participation in the Water and Sanitation Market.* Phnom Penh: GRET.

Hydroconseil. 2013. *Benin: Deep Dive Analysis Report. Global Study for the Expansion of Domestic Private Sector Participation in the Water and Sanitation Market.* Cotonou: Hydroconseil.

Other References

WHO (World Health Organization)/UNICEF (United Nations Children's Fund) Joint Monitoring Program. 2012. "Types of Drinking-Water Sources and Sanitation." Geneva, Switzerland: WHO; New York: UNICEF. http://www.wssinfo.org /definitions-methods/watsan-categories/.

World Development Indicators (database). 2013. Washington, DC: World Bank. http://data.worldbank.org/data-catalog/world-development-indicators.

Appendix B

Table B.1 Types of Improved and Unimproved Sanitation

Type of sanitation	Description
Improved	
Flush toilet	Uses a cistern or holding tank for flushing water and a water seal (a U-shaped pipe below the seat or squatting pan) that prevents the passage of flies and odors. A pour flush toilet uses a water seal but uses water poured by hand for flushing (no cistern is used).
Piped sewer system (sewerage)	Designed to collect human excreta (feces and urine) and wastewater and remove them from the household environment. Sewerage systems consist of facilities for collecting, pumping, treating, and disposing of human excreta and wastewater.
Septic tank	Consists of a water-tight settling tank, which is normally located underground, away from the house or toilet. The treated effluent of a septic tank usually seeps into the ground through a leaching pit. It can also be discharged into a sewerage system.
Flush/pour flush to pit latrine	System that flushes excreta to a hole in the ground or leaching pit (protected, covered).
Ventilated improved pit latrine (VIP)	Dry pit latrine ventilated by a pipe that extends above the latrine roof. The open end of the vent pipe is covered with gauze mesh or fly-proof netting; the inside of the superstructure is kept dark.
Pit latrine with slab	Dry pit latrine that uses a hole in the ground to collect excreta and a squatting slab or platform that is entirely supported on all sides, easy to clean, and raised above the surrounding ground level to prevent surface water from entering the pit. The platform has a squatting hole or is fitted with a seat.
Composting toilet	Dry toilet into which carbon-rich material (vegetable wastes, straw, grass, sawdust, ash) is added to the excreta. Special conditions are maintained to produce inoffensive compost. A composting latrine may or may not have a urine separation device.
Unimproved	
Flush/pour flush to elsewhere	System in which excreta are flushed into the street, yard/plot, open sewer, ditch, drainage way, or elsewhere rather than into a pit, septic tank, or sewer.
Pit latrine without slab	Consists of a hole in the ground without a squatting slab, platform, or seat. An open pit is a rudimentary hole.
Bucket or other container	Container is used to retain feces and sometimes urine and anal cleaning material. Contents are periodically removed for treatment, disposal, or use as fertilizer.
Hanging toilet latrine	Device is built over the sea, river, or other body of water, into which excreta drops directly.
No facility, bush, or field	Excreta are deposited on the ground and covered with a layer of earth (cat method).

Source: WHO/UNICEF 2012.

Table B.2 Demographic, Geographic, and Socioeconomic Indicators for Bangladesh, Indonesia, Peru, and Tanzania, 2010

Item	Bangladesh	Indonesia	Peru	Tanzania
Demographic and area information				
Population (millions)	150	240	48	45
Rural population (millions)	107	120	23	34
Land area (square kilometers)	147,570	1,811,570	1,280,000	885,880
Key socioeconomic indicators				
Per capita gross national income (GNI) (purchasing power parity) (current dollars)	1,649	4,180	8,790	1,410
Gini coefficient	0.32	0.34 (in 2005)	0.48	0.37 (in 2007)
Percentage of population living on less than $2 a day (purchasing power parity)	77	46	13	88 (in 2007)
Percentage of population living below national poverty line	32	13	31	33 (in 2007)

Source: World Development Indicators (database) 2013.

Table B.3 Sanitation Indicators for Bangladesh, Indonesia, Peru, and Tanzania

Indicator	Bangladesh	Indonesia	Peru	Tanzania
Percentage of entire population with improved sanitation (2010)	56	54	71	10
Percentage of rural population with improved sanitation (2010)	55	39	37	7
Number of rural people with access to improved sanitation (2010) (millions)	58.97	46.85	17.77	2.37
Annual loss from lack of sanitation (millions of dollars)	4,200	6,300	—	206

Sources: Data on annual monetary loss from lack of sanitation are from WSP 2013. All other data are from WHO/UNICEF 2013a, 2013b, 2013c, 2013d.
Note: — = not available.

Table B.4 Size, Formality, and Type of Enterprise in Sanitation Sector in Bangladesh, Indonesia, Peru, and Tanzania

Item	Bangladesh	Indonesia	Peru	Tanzania
Employment				
Number of full-time employees	3	3	204	2
Years in operation	11	6	14	9
Years of experience of manager	14	7	12	4
Registration/license				
Percentage of all enterprises	50	9	90	67
Type (percentage of all enterprises)				
Shareholding company	0	6	24	0
Sole proprietorship	97	3	10	70
Partnership	0	0	43	3
Community-based organization	0	0	5	3
Nonprofit organization	3	0	5	0
Not legally constituted	0	91	0	24
Other	0	0	14	0

Note: Totals may not add to 100 percent because of rounding.

Table B.5 Summary Characteristics of Sanitation Enterprises in Bangladesh, Indonesia, and Tanzania

Country/type of enterprise	General characteristics	Estimated total transfers by industry to economy (through labor, finance, tax)	Line of business	Finance and profitability[a]	Outlook toward and view of the poor
Bangladesh					
Prefabricated cement suppliers and builders	Estimated 4,500 enterprises nationally 85% microenterprises 60% have trade licenses Own or rent factory space	$2,700/enterprise × 4,500 enterprises = $12 million	Manufacture and supply of prefabricated cement products; construction and installation More than 50% of revenue comes from sanitation At least 50% of cost is materials	Average capitalization is $5,300 97% cover costs; average margin is 53% (higher in Chittagong and Dhaka region) High level of financing use (multilateral financial institutions, state banks, and so forth)	30% will increase investment in a year; 50% will invest in 3 years linked to easy markets 75% believe low price main motive of the poor; need subsidy and demand creation for quality latrine
Indonesia					
Construction enterprises	Estimated 200 enterprises nationally 70% microenterprises; 30% small 85% not legally constituted (informal single proprietor) but have business license	$13,700/enterprise × 200 enterprises = $3 million	Mainly household installation, but repairs and sale of components also important for business profitability 89% of revenues are from sanitation 80% of cost is materials (little room for innovation)	Average capitalization is $4,800 100% are profitable, average margin is 73% Half of enterprises keep financial records Majority have loans from commercial banks, nonbank financial institutions, and state banks	Optimistic: 80% plan to invest; median investment is $1,000 View the poor as primary client but not certain of their ability to pay even with financing
Pit emptiers	Estimated 750 enterprises nationally 60% microenterprises; 40% small Not legally constituted, but 70% have emptying license	$3,900 × 750 enterprises = $3 million	Use vacuum trucks; districtwide operation 100% of revenues are from sanitation 93% of cost is labor	Average capitalization is $4,600 100% are profitable; average margin is 87% No financial records A few have loans from private banks and nonbank financial institutions	60% plan to invest; median investment is $3,000 Poor are not a primary target

table continues next page

141

Table B.5 Summary Characteristics of Sanitation Enterprises in Bangladesh, Indonesia, and Tanzania *(continued)*

Country/type of enterprise	General characteristics	Estimated total transfers by industry to economy (through labor, finance, tax)	Line of business	Finance and profitability[a]	Outlook toward and view of the poor
Tanzania					
Masons	70% informal, own or rent factory space	$3 × 240,000 units sold per year = $720,000	Mainly construction work; sanitation seen as occasional form of employment; masons engage in other income-generating activities, such as farming 35% of revenues are from sanitation Up to 70% of labor cost is transport San Plat mold a big constraint	Little or no capitalization No financial records	Only 30% plan to expand service range Main target market is the poor
Hardware stores	Legally constituted as sole proprietors	—	Wholesale and retail trade of inputs into construction of on-site sanitation Inventories and transport are largest costs San Plat mold a big constraint	Most enterprises earn profits; large enterprises have margins of more than 50%	Only half have expansion plans Do not target the poor

Note: — = not available.

a. Margins in this column refer to profits as a portion of total revenues.

References

WHO (World Health Organization)/UNICEF (United Nations Children's Fund) Joint Monitoring Program. 2012. "Types of Drinking-Water Sources and Sanitation." http://www.wssinfo.org/definitions-methods/watsan-categories/.

———. 2013a. "Estimates for the Use of Improved Sanitation, Bangladesh." http://www.wssinfo.org/fileadmin/user_upload/resources/BGD.xlsm.

———. 2013b. "Estimates for the Use of Improved Sanitation, Indonesia." http://www.wssinfo.org/fileadmin/user_upload/resources/IND.xlsm.

———. 2013c. "Estimates for the Use of Improved Sanitation, Peru." http://www.wssinfo.org/fileadmin/user_upload/resources/PER.xlsm.

———. 2013d. "Estimates for the Use of Improved Sanitation, Tanzania." http://www.wssinfo.org/fileadmin/user_upload/resources/TZA.xlsm.

World Development Indicators (database). 2013. Washington, DC: World Bank. http://data.worldbank.org/data-catalog/world-development-indicators.

WSP (Water and Sanitation Program). 2013. "Economics of Sanitation Initiative." Washington, DC: World Bank. http://www.wsp.org/content/economic-impacts-sanitation.

Environmental Benefits Statement

The World Bank Group is committed to reducing its environmental footprint. In support of this commitment, the Publishing and Knowledge Division leverages electronic publishing options and print-on-demand technology, which is located in regional hubs worldwide. Together, these initiatives enable print runs to be lowered and shipping distances decreased, resulting in reduced paper consumption, chemical use, greenhouse gas emissions, and waste.

The Publishing and Knowledge Division follows the recommended standards for paper use set by the Green Press Initiative. Whenever possible, books are printed on 50 percent to 100 percent postconsumer recycled paper, and at least 50 percent of the fiber in our book paper is either unbleached or bleached using Totally Chlorine Free (TCF), Processed Chlorine Free (PCF), or Enhanced Elemental Chlorine Free (EECF) processes.

More information about the Bank's environmental philosophy can be found at http://crinfo.worldbank.org/wbcrinfo/node/4.

green press
INITIATIVE

www.ingramcontent.com/pod-product-compliance
Lightning Source LLC
Chambersburg PA
CBHW080614270326
41928CB00016B/3058